To Jacqueline and Alexandra

Scribe Publications
WATT'S THE BUZZ?

Lyn McLean is Australia's foremost consumer advocate on the subject of electromagnetic radiation. After having taught in primary and high schools for many years, she now works full-time on EMR-related issues. She edits *EMRAA News* (the newsletter of the Electromagnetic Radiation Association of Australia), represents the community on a number of national committees, has made submissions to state and federal governments on the health implications of EMR, and has addressed Senate inquiries, conferences, and numerous public meetings on the subject.

In the course of her work, Ms McLean receives many requests from members of the public for information about the health risks of EMR and how they can protect themselves from such risks. She has written *Watt's The Buzz?* to answer such questions in a comprehensive and yet accessible way.

Scribe Publications Pty Ltd
PO Box 287
Carlton North, Victoria, Australia 3054
Email: scribe@bigpond.net.au

First published by Scribe Publications 2002

Cartoon illustrations by Jo Haggie
Typeset in 11 on 14pt Times New Roman by the publisher
Printed and bound in Australia by Griffin Press

National Library of Australia
Cataloguing-in-Publication data

McLean, Lyn.
Watt's the buzz? : understanding and avoiding the risks of electromagnetic radiation.

Bibliography.
Includes index.
ISBN 0 908011 66 0.

1. Electromagnetism - Health aspects. 2. Electromagnetism - Physiological effect. 3. Electromagnetic waves. I. Title.

612.01442

www.scribepub.com.au

Watt's the Buzz?

understanding and avoiding
the risks of electromagnetic radiation

Lyn McLean

Scribe Publications
Melbourne

Contents

Foreword

My son Christopher loved a story at night. He repeatedly asked for a story that had 'something bad happen and a hero'.

Something 'bad' did happen. Christopher was diagnosed with cancer in 1995. For several years my son endured chemotherapy, radiotherapy, bone marrow transplant, surgery, painful injections, blood tests and much more. Despite the treatments, he passed away in 1997, aged seven. He became our hero.

I will probably never know why Christopher contracted this soft-tissue cancer. His bedroom was adjacent to the power box and water heater. He occasionally did talk on the mobile phone, which was a novelty for him. He did peer into the microwave oven to satisfy his ever-present curiosity. Was his illness because of this or was it something else?

Why do our young children continue to contract cancers? Was there something I could have done to save our precious child?

In this book, Lyn McLean has produced a very timely reference work. It is a highly practical work on the topic of EMR from power sources and the telecommunications system. Lyn presents an honest review of the current state of knowledge. More importantly, her book presents some significant but inexpensive ways to avoid unnecessary exposures. Why wouldn't every parent, at least, take the advice?

Is electromagnetic radiation a health hazard? Certainly it is. Microwave ovens will clearly demonstrate that. So will modern military weapons. But what about lower levels of radiation at which telecommunications networks operate and the lower frequencies of the power system? Lyn takes us to all the leading studies. There is no proof beyond reasonable doubt either way. In my view, the jury is still out.

Sadly, industry, and their paid PR people will criticise Lyn for her forthright views. Public health has often been secondary to profits, and the profits from mobile telephones and broadcast media are enormous. On the Australian standard-setting bodies, industry has no time whatsoever for even precautionary measures. I hope that Lyn's book will advance and focus the debate.

My sincere wish is that when all the studies have been completed, the unambiguous outcome is that electromagnetic radiation is not a health hazard—for, if that is not the outcome, we have unleashed a terrible legacy on our younger generation.

For the moment, the practical solutions offered by Lyn will go a long way to ease concerns.

D L Dwyer LL.B

Preface

The fickle finger of fate

I sit at Tullamarine airport happily penning these words, under a host of antennas, surrounded by people talking loudly on mobile phones. I'm on my way home from a hectic two-day meeting in which I've been participating as a community representative at a session devising a code of practice on the siting of telecommunications facilities. This time yesterday I sat in the Melbourne radio studios of the Australian Broadcasting Corporation, awaiting a stint on Sandy McCutcheon's 'Australia Talks Back' national programme. It was my sixth radio interview in a week, and followed hard upon three television interviews.

Why the sudden notoriety?
Last week Sir Richard Doll released a study which found that children who live near high sources of EMR from the power system have double the risk of leukemia. Suddenly the EMR issue is in the spotlight, and suddenly the community's long-held concerns about its health risks are newsworthy.

How did I come to be involved with concerns about the health impacts of EMR, people often ask me? (that is, after their initial question of 'What is EMR?') Is it because I am a doctor of bioelectromagnetism? Is it because I am a terminal activist? Have I lost a child to leukemia? Was I born with a particular gene for tackling environmental Everests?

Well, actually, I reply, it was an accident.
As a teacher, a mother and then director of the Sutherland Shire Environment Centre, I found myself unexpectedly and unwillingly catapulted into the EMR issue in 1996 by the retirement of the centre's resident expert. After much sifting through files, listening, and reading, I was hooked. The question of how EMR affects the body was one of the most fascinating I had ever encountered, and a steep learning curve had begun.

At first my interest was entirely academic. I poured through books, took copious notes, marvelled at the studies, and waded through lengthy reports. This was exciting stuff. This was the cutting edge of science. My left brain was exceedingly happy.

But the fickle finger of fate was not content to engage just half my understanding. I had been exposed to high levels of EMR and I became very sick. Suddenly I could *feel* radiation. Now my world was turned upside down because I could no longer work, no longer tolerate being near mobile phones, no longer tolerate being away from home. Now I understood energy not just with my left brain, but I experienced it with my right. The energy question was my constant companion and the attempt to grapple with it my ever-present challenge.

Feeling EMR is a bit like feeling love. When you're in love you know it and, if you haven't experienced it, you wouldn't have a clue what it's all about. It's an intense and compelling experience. (Othello killed for it.) It's not always accepted or appreciated by others. (The Capulets' insensitivity drove Romeo and his beloved to their deaths). It's not always pleasant (as Macbeth found to his cost), and it changes your life.

EMR has certainly changed my life. My right brain/left brain experience has given me a level of understanding of this issue that I could never have attained by study or illness alone. It has opened windows to areas I'd never imagined, let alone thought of exploring, and it has driven me to examine my values and beliefs in a way that I would not have otherwise done.

It has also taught me, often painfully, what works and what hurts. It is my hope that some of that knowledge will help others to cope with, or better still, to avoid the risks of exposure to EMR. Further, I hope it will be a useful guide to authorities as society negotiates its way through the technological mire that it has created and into a bright new millennium.

Lyn McLean

Acknowledgements

My warm and heartfelt thanks

... to my colleague and merciless critic, John Lincoln, for patiently reviewing draft after draft

... to those who have generously perused the book at various stages of evolution, including Dr. Peter French, Les Dalton, Roger Lamb, Sarah Benson, Diana Crumpler, Peter Llewellyn, and Dan Dwyer.

... to my literary critic, Bob Walshe OAM, whose support has helped me to do what I most love

... to my friends at the Sutherland Shire Environment Centre for their willing assistance in so many vital ways

... to my beautiful daughters, Jacqueline and Alexandra, for sharing the burden as well as the joys

... and to those around Australia—and overseas—who have contacted me about this issue, for trusting me with their experiences.

Can your toaster give you cancer? Is your hot-water heater a leukemia risk?

Every electrical appliance in your home—the toaster, the iron, the washing machine, the fridge, the microwave oven, the hairdryer—not to mention the wiring of your house, produces electromagnetic radiation (EMR). So do computers and mobile phones.

Many people believe that electromagnetic radiation interacts with the body's own energy, causing health problems ranging from headaches and tiredness to cancer and leukemia.

What is electromagnetic radiation? Does it affect people's health and well-being? And, most importantly, how can you reduce your exposure to EMR at home and work?

Read on!

The Context

-1-

How and why?

Odette's story

Eight-year-old Andrew settles himself comfortably into the thick cushions on the lounge-room couch, leafing through the book delivered a week previously by a kindly bearded gentleman in a red fur-trimmed suit and knee-length boots flecked with soot. He is lost in wonder, marveling at the huge armour-plated Stegosaurus, and the flesh-eating giant, Tyrannosaurus Rex, leering from the pages in front of him.

His concentration is soon disturbed by the arrival of his 15-month-old, chubby, auburn-haired sister, Tammy, who bounds noisily into the room, and launches herself in a toddling sort of a way at her brother, burbling meaningfully as she tries to extract the precious book from his grip. Andrew tactfully diverts her attention to the sound of the Wiggles emanating from the stereo on the other side of the room and leans back to enjoy a few minutes of respite as, distracted, she toddles away and begins to gyrate to the tuneful strains of 'Nicky Nacky Nocky Noo'.

Typical though this scene might be of thousands of lounge rooms across the country, it's a scene that, just a few years ago, Andrew's parents despaired of ever experiencing themselves.

When Andrew was four he moved with his parents to a new home in Sydney which, after a substantial facelift, was to be their dream home. It was close to school, shops, and park, had a yard to play in, and was conveniently near a scenic part of one of Sydney's largest rivers.

It was a new beginning, but not of the sort anticipated by his parents. Immediately after moving in, Andrew's mother, Odette, began to experience enormous difficulty sleeping—even though she felt incredibly tired. She also developed severe depression, which was not only out of character but seemed completely unjustified. This was not just an occasional wave of despondency, but

3

a soul-destroying gloom that sapped her energy and interest, and was not relieved when she did drop off to sleep.

Odette also began to develop some worrying physical problems. First was a melanoma, which she had removed. Then came disappointment after disappointment as she and Tony tried unsuccessfully to extend their family. Despite having fallen pregnant with Andrew within four months of trying, Odette was just not able to conceive. But, during a holiday cruise, she and Tony did hit the jackpot. Their joy, however, was short-lived, for not long afterwards they lost the baby. This was the first of three devastating miscarriages. Around the time of the first tragedy, Odette also became seriously ill with pneumonia. She later found that she had developed an antibody for Lupus, an autoimmune disease, which she had acquired since her pregnancy with Andrew.

Odette was not the only one experiencing problems. At the age of six, Andrew had been plagued with health problems all of his young life. He was considerably underweight, had had countless colds and ear infections and, after repeated bouts of tonsillitis, had had his tonsils and also his adenoids removed. He was often ill, and had come close to death with croup. And he had developed the frustrating habit of wriggling down to the middle of his bed every night.

Her husband Tony was not sick in the conventional sense of the word. But he did not experience an abundance of energy, and wondered why it was that his hair stood on end every morning when he woke.

Odette began to suspect that the origin of the family's problems might be the house itself. In a quest for answers, she rang the Electromagnetic Radiation Association of Australia, a public interest group operating from southern Sydney, and she spoke to me.

In a tone of incredible frustration, she described the family's litany of health problems. 'Do you think our problems might be related to the radiation that comes from the wiring?' she asked.

It was certainly possible. 'The fields from the power system are called electromagnetic radiation or EMR for short, and they are associated with a whole range of health problems,' I explained. 'They include some of the problems you've mentioned such as immune problems, miscarriage, difficulties conceiving, tiredness, and depression. There are even reports of children sleeping in high fields wriggling away from the wall the way that Andrew does. It might be worth having the fields in your house measured.' I provided her with information about electromagnetic radiation, and suggested that she contact the Association's chairman, John Lincoln, an electrical engineer, to arrange for an EMR survey.

When John measured the house, he found extremely high fields, particularly in the bedrooms. Odette and Tony's bed was on the on the other side of the wall and just 50 centimetres or so away from the meter box which was emitting a magnetic field of 88 milliGauss (mG) at night when the off-peak hot-water system was operating. The average measurement in the room was around 70 mG, and the lowest reading was 12 mG. Because Andrew's bed was on the other side of the wall from the hot-water system, he was exposed to a field of 12 mG during the entire night.

These readings were well within the guidelines of the National Health and Medical Research Council, which stipulated that people in residences should not be exposed to more than 1,000 mG (5,000–10,000 mG for industry). However, they are far in excess of the 'danger' level of just 3–4 mG that has been indicated by many studies.

John also found high readings in other parts of the house. In the tiny kitchen was the usual assortment of electrical appliances, all emitting fields. Around the digital clock of the electric oven there was a field of 20 mG, and the rear of the refrigerator measured 20 mG. Wherever Odette worked in that room she was sure to be exposed to a field of no less than 20 mG.

On John's advice, Odette and Tony implemented a number of strategies to reduce their exposure. They moved the old hot-water heater out of the laundry, and therefore away from Andrew's bed, and installed a solar hot-water system which they set to heat between 5.00 pm and 7.00 pm so it would not be operating while the family was sleeping. They rectified the problem of an unearthed meter box and, because it was too expensive to relocate the box, Odette and Tony moved their bed to another wall. Odette began washing during the day rather than at night, so that the machine would not be generating fields through the wall against which Andrew was sleeping. She also unplugged the microwave oven when it was not in use to help reduce the ambient field in the kitchen. This reduced the fields to which the family was exposed considerably—although not entirely, with the kitchen still remaining something of a hot spot.

Nevertheless, the family noticed immediate effects. First, there was a period of adjustment, a week in which they felt unsettled and lacked energy. After that, however, their health began to steadily improve. Tony found that his hair no longer stood on end each morning. Andrew improved in health and energy, and no longer wriggled to the centre of the bed at night. To his parent's relief, he gained weight, and friends began to notice the change in his health. And Odette successfully conceived soon afterwards, and produced a healthy baby girl.

Could the high fields in their house have accounted for this family's ill-health? After all, Andrew had experienced health problems ever since he was a baby.

When Andrew was born, his parents were living in a unit in an inner-western suburb of Sydney. Andrew spent long periods of time in a bouncinette against the external wall of the lounge room, a position where Tony also spent many hours regularly. On a recent visit to the units, Odette and Tony noticed that the wall in question contained the power supply for the entire block of flats—four meter boxes! So Andrew and Tony had also been exposed to high fields for long periods at their old address. It may be no coincidence that, while living at that address, Tony had experienced excruciating migraine headaches which have subsequently disappeared.

Odette and Tony's problems were alleviated by reducing their exposure to the EMR from their meter box and household appliances. But sometimes—believe it or not—high fields in the home emanate from the household plumbing.

Dr. Lee's story

Dr. Lee and his family had recently moved into a beautiful brick home in Sydney's Lane Cove. After being there for six months, they learned that the house had a history of serious illness which had touched all of the last four families that had occupied it. A mother and a young child had died from acute leukemia. Another child had developed leukemia while living there, but because she moved away her fate was unknown. A father and a child had also developed gliomas (brain tumours) while living in the house.

John Lincoln found a number of problem areas connected with the wiring in the house. First and most significant, the wiring was earthed to the water pipes and they were conducting an extremely high field—he measured 538 mG at the water meter—through the house. This meant that anyone spending time in part of the house adjacent to the water pipes—working, studying or, especially, sleeping—was being exposed. The water pipes ran not only downstairs but under the upstairs bedroom and up the wall into the adjacent bathroom, so that a person sleeping next to the bathroom wall would be exposed throughout the night.

In addition to the water pipes, the wiring itself was generating high fields. Power entered the house near the upstairs bedroom and ran across the ceiling and down the wall to the meter box, which was on the outside of a bedroom wall downstairs. On the other side of the wall, the field from the meter box measured 45 mG—not a good location for a bed.

Dr. Lee and his family had fortuitously arranged their furniture so that their beds were not in the high fields of the meter box or the water pipes. Nevertheless, John measured fields of 18 mG at one bed, 8.8 to 15 mG at another, and 5.7 mG at a third. While these fields were already too high for comfort, it is likely that they increased further at night when the off-peak electric hot-water system cut in, using substantially more power.

To reduce his family's risk of developing the problems experienced by former occupants of the house, Dr. Lee organised an electrician to install an earth stake and to replace a short length of metal water-pipe with plastic. Not only did this stop the fields in the water pipes, but it meant that the currents traveling into and out of the house through the wiring were now balanced, and so the wiring emitted less of a field. This was a simple but effective solution.

Two powerline stories

Naturally, homes situated close to high-voltage powerlines are particularly vulnerable.

Murray and his wife Susan wanted an EMR survey done of a property they were interested in buying. It was a beautiful home with a beautiful aspect, situated on a bushy peninsula that jutted into a particularly scenic part of Sydney's Georges River. However, there was a potential problem. A high-voltage powerline ran right through the property before eventually spanning the river. In the backyard, the space inside one of its pylons had been utilised to accommodate the clothesline! The line ran along the entire side of the house, around ten metres away from all of the bedrooms. Inside the house, in the rooms adjacent to the powerline, John Lincoln measured magnetic fields of between 8.4 and 10 mG during the day. On the side of the house furthest from the powerlines he measured an average of 5 mG. These fields were likely to be very much higher during the evening when the line was conducting more current to feed the usual demand for appliance use.

John provided the couple with the readings, and explained the risks of fields at these levels that have been identified in the research. Murray and his wife went ahead with the purchase, and soon moved in with their young children.

Later, John learned in a conversation with the estate agent that the son of the previous owner of the house had died of leukemia.

On another occasion, John measured a home owned by Mathew and Stella close to a high-voltage powerline in Sydney's west. Inside the house, fields were generally in the order of 12 mG. In the course of discussion, John remarked that the couple seemed to be in good health.

'We're concerned about our health and we take pretty good care of ourselves,' replied Mathew. 'But I wouldn't have my grandchildren living here for quids.'

'Why not?' asked John, surprised.

'Because the neighbours on each side of us have both lost children to leukemia.'

Kate's story

Sometimes problems are caused, not by the magnetic fields, but by the *electric field* that is also a component of EMR.

Kate, a secretary, rang me, desperate to find a solution to a string of difficulties she had been battling for several years. First, she had severe dermatitis on her hands which had persisted with barely a let-up for over three years, and made even simple tasks painfully difficult. Second, she was plagued by incredible tiredness which made functioning after 6.00 pm almost impossible, to the point where she was often unable to string coherent sentences together. Even though she was constantly tired, Kate had a great deal of difficulty sleeping, and the quality of her sleep was poor.

By the time she spoke to me, Kate's difficulties while working at the computer had increased to the point where they had become intolerable. She now experienced a burning sensation when using the mouse. Moreover, the burning sensation travelled up her arm, persisted overnight, and was only alleviated by her keeping away from the computer for days on end—which was quite difficult, given that her livelihood depended on computer use. She also began to develop pain behind her right eye, and the condition of her hands deteriorated to the point where they had constant open cuts which bled easily on impact with just about anything.

When John Lincoln visited Kate's workplace, he found extremely high electric fields. The highest were from the keyboard, itself, which measured up to 300 volts per meter, whereas most keyboards that John has measured were about five volts per metre. (Though this field was substantially smaller than the 5,000 volts per metre allowed by the Guidelines of the National Health and Medical Research Council, it was far above the 10–40 volts per meter that have been implicated by one study with leukemia.) High electric fields were also coming from wiring that ran under the floor beneath Kate's workstation (56 volts per metre), and inside the wall at the rear of her desk.

In this case, the solution was simple. With the cooperation of Kate's employer, John moved the computer to an alternative location, and found that the fields immediately dropped from 300 to just five volts per metre. Within a week, Kate's dermatitis had completely and permanently cleared, and her tiredness greatly abated.

Ron's story

Not all health problems from EMR are caused by the 50 Hz power system. Ron was a successful businessman who often conducted his work from overseas locations as far apart as Singapore and Paris. Needless to say, he was an extremely heavy mobile-phone user, making calls that lasted sometimes for an hour, and running up bills of around $1,000 per month. He invariably held the phone in his left hand so that he was free to work with his right. Often during calls he experienced a sensation of heat about two or three centimetres *inside* his head near the handset, and this feeling often persisted for up to twenty minutes after a call.

In 1997 Ron was diagnosed with a Merkel Cell tumour on the left side of his forehead at the very point where the aerial of his phone had touched his head. Merkel Cell tumours usually start in the lungs, from where they spread to other parts of the body. However, there was no indication of a tumour in Ron's lungs. Something other than the normal course of events had occurred in his case. The tumour was surgically removed.

The following year Ron was diagnosed with a tumour in the parotid gland, just below his ear on his neck and shoulders, directly and precisely adjacent to the position of the mouthpiece of his mobile phone. This was a very aggressive tumour that required radiotherapy and radical surgery, which removed a sizeable chunk of the left side of his neck and shoulder. Ron was told he had just two more years to live.

Thankfully, Ron has now outlived the forecast time limit, but he bears and will continue to bear significant scars, the physical and emotional legacy of his whirlwind romance with technology.

For Ron has no doubt whatsoever that his tumours were caused by the radiation from his mobile phone. The match between the position of his mobile phone and the location of the tumours was too perfect to be a coincidence.

And so it often is. The connection between high levels of EMR and health problems is just too frequent, too compelling, to be merely coincidence.

Time and again, John Lincoln and I find that people living or working near high fields develop health problems. In our experience, they are exhausted and they are unwell, sometimes hugely so. In our experience, they are immensely frustrated at the difficulties in finding a solution to their problems. Fortunately, in our experience, many of their problems can be solved by reducing their exposure.

As it happens, our experiences of health problems from EMR are not isolated incidents shrouded in a nebulous mist of hypothetical conjecture. They are backed by a solid and growing body of scientific evidence that exposure to EMR is indeed linked with a whole range of health problems.

What these problems are and how to avoid or reduce them is the subject of the remainder of this book.

The Basics

-2-

Energy fundamentals for absolute beginners

W e are surrounded by electromagnetic radiation. The sun emits EMR, which we experience as light and heat. The earth has a static magnetic field that regulates our body cycles and helps living creatures find their way. Even people emit electric and magnetic fields which can be measured, for example, with a magnetencephalogram and an electroencephalogram.

To the naturally occurring electromagnetic radiation to which our bodies have adapted over eons, man has added a host of artificial emissions. In the last one hundred years, society has witnessed the proliferation of huge power-distribution networks, the advent of radio and television, the widespread use of computers in industry and homes and, more recently, the emergence of a huge and constantly evolving telecommunications network. If EMR were visible, we would see ourselves enveloped in a complex web of tightly woven emissions.

To understand the impact of EMR on our bodies it is useful to understand something of the nature of this energy. The following description is a fairly simple outline that will be useful in understanding some of the concepts used elsewhere in this book. Don't be alarmed. It's entirely suitable for the science dropout and the technically incompetent!

The electromagnetic spectrum

EMR is classified according to its frequency (the number of wavelengths to pass a given point in one second), which is measured in Hertz (Hz). Our power system operates at a frequency of 50 Hz (60 Hz in the US); GSM mobile phones operate at frequencies of between 890 and 915 MHz (million Hertz); third-generation mobile phones operate at around 1800 MHz; and microwave ovens operate at the still higher frequency of 2,450 MHz.

Looking at the following picture of the spectrum, you can see that EMR ranges from extra low frequency (ELF), through the radiofrequency band in which our telecommunications system, microwave ovens, radio, and television operate, up to infrared and visible light. This part of the spectrum is known as non-ionising radiation because the energy here is insufficient to break covalent chemical bonds directly. However, that does not mean that this energy is inert. There are other types of chemical bonds, and other less direct ways in which energy can interact with them. Beyond the non-ionising part of the spectrum is ionising radiation, which includes ultraviolet light, x-rays, gamma rays of the nuclear industry, and cosmic rays—which are known to cause health problems and are beyond the scope of this book.

The energy in each part of the non-ionising spectrum is known by a particular name—Extra-Low Frequency (ELF of the power system), radiofrequency radiation (RF of mobile communications technology), or microwave radiation. However, for the purpose of simplicity, the term *electromagnetic radiation* (EMR) will be used in this book to apply to all of these emissions.

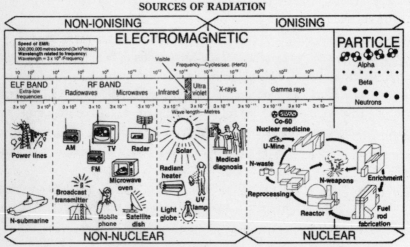

A chart showing how electromagnetic and particle radiations are associated with various medical, telecommunications, nuclear and other activities.

Reprinted with permission from Dalton 1991.

Electromagnetic radiation can be considered as two fields: the electric field and magnetic field.

Voltage and current

Without going into a complicated dissertation on physics, to understand an electric field it is necessary to first understand the concepts of voltage and current. Think of a bucket of water with a hole at the bottom. The force (voltage) of the water causes it to escape from the hole in a stream (current). If the force is great the water will gush from the bucket; if it is not, the water will just trickle out. If the force is not steady but varies, so does the stream of water. So it is with voltage and current.

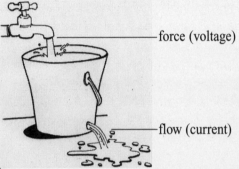

Electric fields

An electric field is the result of voltage (force) and is present whenever an appliance is turned on. Your coffee percolator, for example, will be giving out an electric field as it prepares its tantalising brew first thing in the morning.

What many people don't realise is that the electric field is also present when an appliance is plugged in but switched off. So that coffee percolator is likely to be generating a field throughout the day. Let's return to the bucket analogy for a moment. When the hole is plugged, water cannot gush out as it did before, but the pressure forces minute amounts of water to escape from any imperfectly sealed holes in the bucket. In the same way, when the appliance is turned off at its switch, a small electric field 'escapes' from the wire. Just as a greater force of water will cause more leakage, a higher voltage will result in a higher electric field.

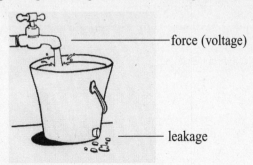

Leakage (electric field) through imperfectly sealed hole

This means, of course, that the only way to eliminate the electric field from the wiring up to an appliance is to remove the plug from its socket.

The tendency for electric fields to 'escape' as we have seen above can be experienced on a rainy day near high-voltage powerlines. That crackling sound you will hear is from the corona, and is the sound made by minute amounts of current as they escape from the wire through the moist air.

Electric fields can, to some extent, be blocked by solid objects that are connected to earth; or, more accurately, these objects simply divert the energy back to earth. Buildings, trees, and furniture all provide some protection. The fields are most effectively diverted with earthed sheets of metal, and these can be used in the home to block the electric field from a meter box, for example. They can also be blocked with insulation such as rubber or PVC, which is why wiring is encased in PVC conduit.

Electric fields are measured in volts per metre (V/m) or variations of this, and occur at right angles to the conducting wire as in the picture below.

Magnetic fields

Whenever electrons are flowing they create a current, and this produces a magnetic field. So an appliance must be turned on at the powerpoint and operating in order to generate a magnetic field. This means that your coffee percolator is generating both an electric and magnetic field while brewing the morning pick-me-up, but only an electric field while it is idle during the day.

In general, the greater the current, the higher the magnetic field will be.

Unlike electric fields, magnetic fields are difficult to block. They pass through the human body, most objects, and even the earth. Powerlines that have been laid in underground trenches can, depending on the way they are laid, still produce a measurable magnetic field on the nature strip.

Magnetic fields are measured in gauss (G), milliGauss (mG) or, alternately, tesla (T) and microTesla (μT). They exist around the conductor as in the diagram at the foot of the oppposite page.

The strengths of the electric and magnetic fields diminish with distance from the source, so that the further away you are, the less the effect of the fields.

Electromagnetic waves

The 50 Hz signal that is produced by the power station (60 Hz in the US) is a pure, gently undulating wave, known as a sine wave, which can be depicted on a graph like this:

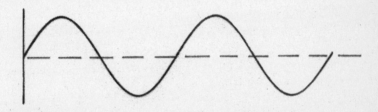

As you can see from the diagram, alternating waves are quite symmetrical, with half being positive and the other half negative. This is the sort of signal that most laboratory animals and cells are exposed to by scientists attempting to ascertain the safety—or otherwise—of EMR from powerlines.

However, the signals to which people are exposed in real life are quite different.

Modulation

A modulated wave is one in which the sine wave carries another wave superimposed on it. Effectively, this means that two quite separate frequencies are traveling together either along a powerline or through our airways. Usually, modulation occurs deliberately, in order to convey information. For example, in many areas, powerlines are used to convey information about electricity consumption so that workmen no longer have to manually check individual electricity meters and brave ferocious guard dogs. Often, however, modulation is unintentional and occurs when powerlines, particularly high-voltage lines, act as antennas for a telecommunications system. (Vignati and Giuliani 1997, Lundquist 1997.) In this situation, the powerline acts like a receiver/transmitter, albeit an inefficient one. Sometimes additional frequencies are introduced by transformers and switches in the system.

Amplitude Modulated (AM)

The operation of our radio networks relies on two different systems of modulation, AM and FM. The AM system is amplitude modulated, which means that the signal has the same frequency or regularity, but a different amplitude or size.

The FM system is frequency modulated, so that the amplitude or size is constant, but frequency or regularity varies. This occurs when the wave is compressed or stretched so that there are more or less waves than the usual number of the carrier signal. In an FM system, the receiving system interprets this complex signal and uses it to reconstruct the original transmission exactly. Consequently, it allows a complicated signal to be transmitted with fewer errors—and therefore more clarity—than does the AM system.

Frequency Modulated (FM)

Another type of modulation occurs when the power of a transmitter or power current is switched on and off quickly, as occurs with a strobe light. This creates both radiofrequency and low-frequency signals.

Transients

Transients are a type of unwanted modulation that are caused by motors, switches, and electrical devices being switched on and off throughout the power grid. A spike is a type of transient, rather like a pimple on the sine wave, which lasts only for a single cycle or less, and is followed by smaller reverberations. (This can be compared to the action of a diving board, which dips low and springs high during a dive and bounces several times afterwards.) So, by the time the 50Hz signal reaches us, it looks more like the image below.

These tiny spikes and transients are effectively a high-frequency signal that is now superimposed on the original sine wave. This is a far cry from the signal that was originally emitted from the power station—and a far cry from what is often tested in laboratories.

Thyristors

Transients are often produced by the use of thyristors. These are electronic switches that enable appliances such as drills, electric mixers, and hairdryers to run at variable speeds.

Turning on or off an appliance containing a thyristor causes a distortion of the wave through the power system.

Transients, spikes, amplitude modulations, frequency modulations, and other 'hash' signals all combine to produce a wave that is substantially different to the one that left the power station—and one to which luckless laboratory mice are subjected.

Outside the power system, the telecommunications network operates at a higher frequency, which produces waves of shorter lengths. Whereas the old analogue mobile-phone system operated using a steady sine wave, the new digital network utilises sharp pulses of power so that the signal looks a little like this:

Pulse Modulated (PM)

However, as in the power system, the signals are not even, but consist of a series of tiny peaks and troughs caused by mobile phones switching on or off, for example. This is extremely important because it means that our bodies, once again, are not being subjected to an even or regular pattern, but a series of jarring signals of varying intensities.

In constructing standards for radiofrequency protection, there has been a tendency to estimate the impact on the body by averaging out the signal over six or so minutes rather than considering the effects of brief but intense bursts of energy. This may be rather like claiming that a bullet to the heart is not harmful because, averaged over six minutes, it would cause nothing more than a slight bruise.

Now that you have a better understanding of the nature of electromagnetic waves themselves, it's time to see how they are applied in our power and telecommunications systems.

-3-

The power network

t drives industry, lights our homes, powers a host of labour-saving devices, and enchants us with a tantalising assortment of entertainment. Electricity: it's as familiar as our morning cup of coffee, and so we expect it to be safe.

But it may not be safe, at least for some people in the community.

There is increasing evidence that the power network brings—along with convenience—an element of risk from the radiation it emits. This radiation comes principally from the powerlines and substations that supply our homes and businesses, but it also emanates from internal wiring and every electrical appliance we use.

Fortunately, however, we don't have to dispense with the entire electricity grid to protect our family's health. By avoiding sources of high fields and judiciously shielding them, it is possible to reduce the 50 Hz fields—and with them the health risks—in most homes and workplaces, often quite substantially.

To understand how you can reduce your exposure from powerline sources, it is useful to have a basic understanding of how the network operates.

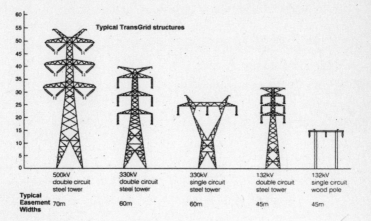

Illustration courtesy of Transgrid

21

Electricity is generated at a power station and transported by high-tension wires—at between 110 and 500 kV—to local centres. Then it is transformed to lower voltages at a series of substations till it reaches the 240 volts of our domestic power supply.

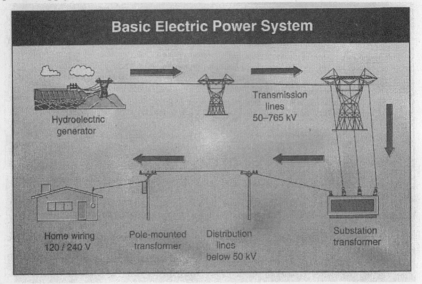

Illustration courtesy National Institute of Environmental Health Sciences, National Institutes of Health

Domestic power poles carry four wires. Of these, there are three active wires which distribute electricity to homes and workplaces (this is called a three-phase system). The three active phases conduct electrons through the power grid. When they have imparted their energy to power appliances, they are returned to the power station via the neutral wire to be re-energised. Conducting the return or neutral current back to the substation is the job of the fourth wire of the domestic power system.

The amount of radiation you receive from a powerline will depend on the spacing of these phases and, of course, how far away you are from them. If the active wire and its corresponding neutral wire are close together, the fields cancel each other out to some extent. Wires that are bundled (grouped closely together, and sometimes twisted around each other) produce a greatly reduced field.

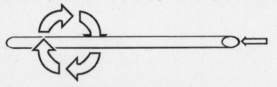

Magnetic field around active wire

Magnetic field around return wire

In wires close together the fields cancel each other out

In an underground distribution system, cables are shielded by the earth and so produce almost no electrical field. The magnetic field, however, which is hard to shield, is radiated through the earth and can produce a measurable alternating field at ground level. The magnitude of this field will depend on how the cables have been laid. Most commonly, the three active phases and the neutral are laid in separate conduits. This is not as effective in reducing fields as laying three-phase cable and neutral in a single conduit because, as we saw above, when wires are bundled closely together their fields cancel each other out.

The neighbourhood distribution system includes strategically placed transformers which may be either mounted on poles or free standing. Their function is to reduce the voltage from high-voltage powerlines to the 240 volts at which our domestic system works.

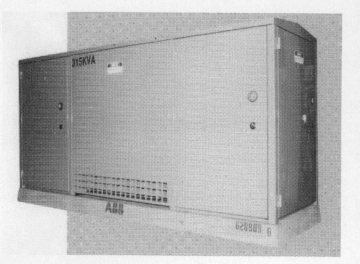

Finally, electricity is conducted from one of the active phases of the power-lines to the service point which is the beginning of the home wiring system. From there it flows to the meter box and then to the fuse box which contains fuses (or circuit breakers) for different circuits within the house. In an average home there are separate fuses and circuits for power outlets, lights, the electric hot-water system, and the electric stove. This separation reduces the amount of current flowing through any one wire, in order to lessen the danger of fire from over-heating. The wiring flowing through the house consists of one active, one neutral, and one earth wire often connected to the water pipe.

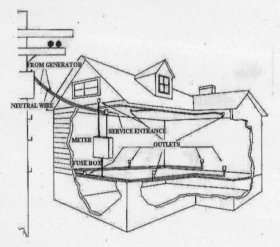

Illustration courtesy of Integral Energy

Inside the home, the wiring configuration can produce quite a difference in the fields that are generated. As mentioned above, locating active and neutral wires close together or in a bundled arrangement reduces the fields, as does the use of shielding. In an experiment, John Lincoln (the managing director of EMR Surveys) ran one active and one neutral wire separately, and measured a field of 300 mG. The same wires moved closer together produced a field of 30 mG. When twisted, the field they emitted was just 2.1 mG. Obviously, twisted cabling will produce the lowest emissions in your home.

As we have seen, electric current is not consumed by use, like water, but needs to return to its point of origin. This means that the electrons that have powered our appliances continue to flow through the wiring, and return (via the return wire) to the service point, then through the power grid and, ultimately, to the power station.

However, not all of the current distributed by the electricity grid is returned to the power station via the powerlines. Because all metal objects in a home must be earthed for safety, some current returns 'to earth', through the earth wire, via metal earthing stakes and thence through the ground to the power station. So the earth itself is something of a 'sink' for the distribution system.

In many homes the earth wire is connected to the neutral terminal and bonded to the water pipes because these provide a good connection to earth. When this occurs, the neutral current may travel through the water pipes, taking whatever route it can find back to the power station. In a house that has been earthed in this manner, currents in the water pipes can occasionally generate quite high fields under the house; these can cause problems for the occupants, particularly if a bed or workstation is located over the pipes. If necessary, current can be eliminated from the water pipes by asking a plumber to insert a section of non-conducting plastic pipe (see chapter ten).

Does radiation from the power network pose a health risk?

Countless studies during more than three decades have indicated that this is indeed the case. From the ground-breaking 1979 study by Nancy Wertheimer on children living in high-current situations, there has been strong evidence of an association with childhood leukemia; this remains, perhaps, the best-researched association. But fields from the power system have also been associated with a range of other complaints, including brain tumours, cancer, depression, Alzheimer's disease, and low immunity, as you will see in chapter five.

Well-known epidemiologist Dr. Samuel Milham has researched the connection between the prevalence of childhood leukemia and the installation of electrical networks. He found that a peak occurred in childhood leukemia in children aged between two and four just after the electrification of their neighbourhoods. He also found that, in many countries, the rate of leukemias was proportional to the extent of electrification. He concluded that, 'The childhood leukemia peak of common acute lymphoblastic leukemia may be attributable to electrification', and that at least 60 per cent of childhood leukemia is preventable. (Milham and Ossiander 2001.)

In March 2001 the safety of the power network became a focus of media interest when well-known epidemiologist Sir Richard Doll (the man who had made the connection between smoking and lung cancer some decades previously) released the results of his latest research. In a report for Britain's National Radiation Protection Board (NRPB), the advisory group of which Doll was chairman reviewed a number of previous studies on powerlines and leukemia. It found a doubling of the risk of childhood leukemia at exposures of 4 mG and over.

While the NRPB and electricity utilities in both the UK and Australia were quick to downplay the significance of the findings, intense media focus helped generate a response from authorities in Australia. The federal health minister Dr. Michael Wooldridge 'warned families living close to high-voltage powerlines to keep very young children away from the potentially cancer-causing electromagnetic radiation.' (*Sydney Morning Herald* 7 March 2001.)

Dr. John Loy, chief executive officer of the Australian Radiation Protection and Nuclear Safety Agency (ARPANSA) said that 'there is no formal exposure standard and the current health guidelines for exposure to ELF fields were released in 1989 ... the rationale for the exposure levels is dated. I will ask the RHC [Radiation Health Committee] to place review of the interim guidelines and the need for an exposure standard on its agenda for early action. The issue of how such a standard might deal with precautionary approaches to limiting exposures of children clearly will be important.' (ARPANSA release 15 March 2001.) Later Dr. Loy advised me that the Radiation Health Committee had agreed to the development of a standard for power exposure, and would undertake that process after completing the standard for radiofrequency radiation exposure.

This is an important step, as Australia's existing guidelines are clearly dated and inappropriate.

What levels of exposure are safe for powerlines?

In Australia there is no official standard for the amount of exposure that people can receive from powerline sources. In 1989 the National Health and Medical Research Council of Australia (NHMRC) adopted guidelines established by the International Radiation Protection Association (IRPA). These guidelines permit people in residences to be continuously exposed to a magnetic field of 1,000 mG and an electric field of 5 kV/m (5,000 volts/metre). They permit people in industry to be continuously exposed to a magnetic field of 5,000 mG and an electric field of 10 kV/m (10,000 volts/metre). (NHMRC 1989.) These are not science-based health standards.

Allowable exposures	NHMRC Guidelines	
General Public	**Magnetic Fields**	**Electric Fields**
Whole working day	1,000 mG	5,000 V/m
Few hours per day	10,000 mG	10,000 V/m
Occupational		
Whole working day	5,000 mG	10,000 V/m
Few hours per day	50,000 mG	30,000 V/m

These levels of exposure are *far* above those at which many health problems have been detected. Over the past few decades a number of studies have found that health problems occur at exposures to magnetic fields of just 3–4 mG (compared to the 1,000 mG allowable in the NHMRC Guidelines;); in some cases, health problems occur at lower levels. Here is a summary of the findings of some of those studies, which are elaborated upon in chapter five:

- Feychting and Ahlbom 1993 (leukemia risk at 1 mG and above)
- Green 1999 (leukemia risk at 1.4 mG and above)
- Greenland 1999 (leukemia risk at 2 mG and above)
- Feychting 1997 (leukemia risk at 2 mG and above)
- Michaelis 1997 (leukemia risk at 2 mG and above)
- Linet 1997 (leukemia risk at 3 mG or above)
- Tomenius 1986 (cancer risk at 3 mG or above)
- Savitz 1988 (cancer risk at 2 mG or above)
- Feychting 1998 (risk of dementia at 2 mG or above)
- Ahlbom 2000 (doubled risk of leukemia at 4 mG)
- Doll 2001 (doubled risk of leukemia at 4 mG)
- Schüz 2001 (increased risk of leukemia at 2 mG)

While it may not have any official status, the 2–4 mG danger limit has been recognised by a number of authorities. In 1990 the US Environmental Protection Agency (EPA) produced a report which concluded that, 'The strongest evidence that there is an association of certain cancers (namely leukemia, cancer of the CNS and lymphoma) with exposure to magnetic fields comes from the childhood cancer studies. Several studies have consistently found somewhat elevated statistically significant risks and elevated non-significant risks of these three site-specific cancers in children whose exposure to magnetic fields has been estimated by the types of wires near their home or magnetic field measurements of 2 mG or higher.' (US EPA 1990.)

Subsequently, the Victorian Department of Health commissioned a report on the effects of EMR from power sources, which was conducted by Melbourne University. The report concluded that there was 'prima facie evidence that a residential exposure to powerline frequency magnetic fields of at least 3 mG is associated with an elevated risk of childhood cancer.' (UMSCC 1990.)

Unfortunately, there are not many studies that have evaluated the risk of electric fields, so it is somewhat harder to arrive at a 'safe' level of exposure. However, one study that does suggest a risk level was conducted recently in Canada. In their study of over 31,000 workers at Ontario Hydro, Paul Villeneuve, and Anthony Miller found a higher rate of leukemia and Non-Hodgkins Lymphoma particularly among workers exposed to fields between 10–40 V/m. Those who had worked in high fields for the longest time had eight to ten times the risk of leukemia. (Villeneuve et al. 2000.)

As you can see, these levels are significantly lower than the 5,000–10,000 volts/metre allowed by the NHMRC Guidelines.

It is certainly my observation that exposures to electromagnetic fields that are just a fraction of the allowable level are associated with health problems. Many callers ring concerned about whether high fields from power sources might be the cause of their family's health problems. Often, on inspection, high fields are found in the house—many caused by water pipes conducting current—and, when reduced, symptoms generally abate.

In the course of conducting EMR surveys over a number of years, John Lincoln has heard some interesting anecdotes about the health/exposure connection. In several situations where high-tension powerlines produced fields of 11 or 12 mG, residents commented that a neighbour—or previous occupant—had developed leukemia. On one occasion, a real estate agent watching an EMR survey expressed his belief that powerlines contributed to health problems. When asked why, he explained, 'In my first six years in real estate I sold new houses under some large powerlines, and almost always they would be resold within two years. Families who would give a reason all said that they had had nothing but sickness since moving in.'

Such anecdotes may not have the force of empirical scientific research, but they do support the precautionary value of limiting exposure as much as possible.

To what are people being exposed?

People are exposed to electromagnetic radiation from a huge variety of sources. Not only does our domestic power-supply produce fields, but so do household appliances, office equipment, lighting, cars, trains, and aeroplanes.

A German study measured the exposure of nearly 2,000 volunteers over a 24-hour period between 1996 and 1997. The researchers found that the average was 1 mG. A marginal difference could be seen between summer and winter, and highest exposures were found to occur at work. (Brix et al. 1999.)

Of course, the level of radiation to which you are exposed can vary greatly according to your occupation and location. Dr. Luciano Zaffanella surveyed 1,000 people in the United States to ascertain their field exposure in a typical day. He found that:

- about 3 per cent were exposed to over 2 mG at school;
- over 26 per cent spent more than one hour in fields over 4 mG;
- over 9 per cent spent more than one hour in fields over 8 mG;
- about 1.6 per cent were exposed to at least 1,000 mG;
- highest exposures occurred at work, especially among electrical and service

workers including cooks, housekeepers, police, prison guards, and waiters; and
- lowest exposures occurred at night while sleeping.

(Zaffanella 1998)

Is it safe to live near powerlines?

Many people are concerned about the amount of exposure they are getting from powerlines outside their home, or wonder whether to purchase a new home near powerlines. Many believe that the metal towers that support the lines are a risk to their health. However, it is the lines that emit electromagnetic radiation—not the towers, which merely support the lines. In fact, there are often higher readings from the lines in the middle of a span, where they are closer to the ground, than at the towers, where they are higher and therefore further away from us.

In the United States, the EPA compiled data on exposures from a range of high-voltage powerlines, which gives a rough indication of exposures at various distances.

	Beneath (15.24m)	50' (30.48m)	100' (60.96m)	200' (91.4m)	300'
115 kV line					
average use	30 mG	7 mG	2 mG	0.4 mG	0.2 mG
peak use	63 mG	14 mG	4 mG	0.9 mG	0.4 mG
230 kV line					
average use	58 mG	20 mG	7 mG	1.8 mG	0.8 mG
peak use	118 mG	40 mG	15 mG	3.6 mG	1.6 mG
500 kV line					
average use	87 mG	29 mG	13 mG	3.2 mG	1.4 mG
peak use	183 mG	62 mG	27 mG	6.7 mG	3.0 mG

(US EPA 1992)

Distance from powerlines is not the only factor you need to consider to ascertain whether or not your family is at risk from the radiation from high-voltage powerlines. As you can see from the table above, the magnetic field varies considerably according to the demand for electricity, which affects the amount of current flowing through the line at any one time.

It can also be misleading to consider your exposure to the radiation from powerlines without also considering what fields are being generated from sources within the house. Sharon rang John Lincoln to organise an EMR survey of her home because she was worried that the high-voltage powerlines next to the house might be contributing to the health problems that she and her husband were experiencing. Sharon was suffering from a rare type of lymphoma, her husband had Alzheimer's disease, and both of them had a history of sickness spanning many years. John found that there were indeed high fields within the house—but not from the powerlines. The water pipes were radiating fields of 145 mG under the couple's bed. Fortunately, fields from water pipes can be eliminated, whereas those from high-voltage powerlines cannot.

Is it safe to live near a substation?

Many people are concerned about whether a substation emits fields that are dangerous to health. This will depend, to some extent, on the size, construction, and location of the building. Older substations are often bigger than their more modern counterparts so that the fields at the perimeter are lower. Newer substations are not only more compact, but they tend to be located much closer to homes and workplaces, and are sometimes located in the basements of blocks of units even though they appear to be better shielded.

Even if homes near a substation are not affected by the radiation emanating from the substation itself, they can be affected by the powerlines surrounding it. Not only do many powerlines converge at a substation, but they may be closer to the ground so that fields at ground level would be higher. Moreover, there is likely to be a higher than average level of return currents in the earth surrounding substations.

Is it safe to live/work near a transformer?

All transformers tend to emit high electromagnetic fields. Those that exist as part of the power system are encased in a steel box that provides shielding to reduce the fields somewhat. Nevertheless, the large transformers found on our streets can emit around 50 mG.

In addition to the transformers that exist as part of the power grid, some transformers are located in large buildings. They convert the high-voltage power that is often transported in underground lines as 3.3 kV or 11 kV to 240 volts necessary for operation of equipment within the building. Often at the same location is a profusion of electrical equipment such as switchgear, circuit breakers, and switch boards. Together these produce high electromagnetic fields which permeate the walls of the building and can affect the health of residents or workers in neighbouring rooms. In the course of measuring the fields in many units and workplaces, John Lincoln has found fields from transformers and associated equipment measuring up to 700 mG.

In the United States Dr. Samuel Milham conducted a study of a group of workers in an office situated above three 12 kV transformers. Fields were 190 mG at floor level and 90 mG at just over a meter above it. He found that over a fifteen-year period there were eight cases of cancer. Only one occurred among people who had worked in the office for less than two years, and the other seven were found in the 156 people who had worked there for two or more years. The risk, he found, increased with the length of employment. (Milham 1996.)

EMR surveys

The most accurate way of ascertaining your family's exposure is to have the fields in your home measured. An EMR survey will determine whether or not there are high fields from sources inside or outside the home, and will provide advice about how to reduce fields where possible.

Some surveys will provide information only about the magnetic fields within the house. This is not always helpful and can even be misleading, in that magnetic fields vary during the day according to how much electricity is being used in the home (or neighbourhood) at the time. Magnetic fields measured in a residential area during the middle of a mild autumn day when very few appliances are being used in the area are often quite low. However, during the evening when families are cooking their evening meals and using their heaters, or during the night when offpeak hot-water services are operating, magnetic fields will generally be higher.

For this reason, an EMR survey should ideally also take measurements of electrical fields. Because these are not affected by the amount of electricity being used, they are fairly consistent throughout the day.

Where there are indications of problems, it is sometimes helpful to take overnight readings of magnetic fields. John Lincoln uses a data logger, which takes measurements of electric and magnetic fields at regular intervals during the night and provides information in graphic form via a computer program.

What are the risks to my property value?

It is not just health that is at risk from living close to a high-tension powerline or substation. As many people have found to their cost (literally), this infrastructure can adversely affect the value of their properties.

Our agency believes it's a really great buy, Mrs Smith. It's $50 000 less than anything comparable elsewhere.

Around the world there have been a number of examples which show that powerlines do devalue property:

- In 1985 a Texan court awarded damages of $104,275 to a school district after an electrical company constructed a high-tension powerline 130 feet and 250 feet respectively from two schools. The line was later moved at a cost of $8 million. (Brodeur 1989.)
- Peter Colwell, writing in the *Journal of Real Estate Research* (US), said that property values were reduced near the metal towers that supported high-voltage powerlines. (Colwell 1990.)
- Robin Gregory and Detlof von Winterfeldt also found reductions in the value of real estate values near high-voltage powerlines. (von Winterfeldt 1996.)
- In 1993 a New York court ruled that—when a landowner's property is seized for the construction of high-voltage powerlines—the landowner was to be compensated for the loss of value of the remainder of his property due to 'cancerphobia', whether or not that fear is justified. (*New York State Court of Appeals* 12 October 1993.)
- Also in 1993 a study by a Houston appraiser found that the value of ten properties adjacent to a high-tension line was 13 to 30 per cent lower than those of comparable homes nearby. (*Wall Street Journal* 8 December 1993.)
- British building societies are reportedly refusing mortgages for homes under powerlines in response to health concerns, with the effect that these homes have been devalued by up to 33 per cent. (*Birmingham Post* 25 September 1995.)

- A US court ordered Virginia Electric and Power Co. to pay $967,000 to property owners John and Janet Dolzer for devaluation of their property by EMFs from two 230 kV transmission lines. (*Microwave News* March/April 1996.)
- A Canadian court ordered that fields from a substation devalued an adjacent property, justifying a reduction in tax for the owner of several thousand dollars annually. (*Microwave News* March/April 1997.)
- A US bank refused to mortgage a property next to a substation on the basis that it might cause the property 'to suffer from environmental conditions' (that is, according to a bank representative, it was devalued by $50,000 because of its location). (*Microwave News* September/October 1998.)

In Australia
- A Melbourne real estate agent has claimed that greater awareness about EMR has influenced approximately 90 per cent of buyers to reject houses near powerlines, whereas ten years previously the lines were no such deterrent. (*Age* 17 April 1996.)
- A real estate agent from Moonee Valley said that homes near transmission lines could be up to $20,000 cheaper than comparable homes elsewhere. While his agency did not have trouble selling these homes, buyers tended to rent them out rather than live in them. (*Herald Sun* 3 April 2001.)
- I am frequently contacted by prospective home buyers to ascertain the safety of properties near high-tension powerlines. Most report that the home in question is extremely cheap by usual standards.

-4-

The telecommunications network

I t's sleek, colourful, compact, and convenient. It's your umbilical cord to society. With it you can be instantly available to friends, you can work outside the office, you can access the internet and, as industry so often reminds us, you can call for help in emergencies. The mobile phone, in all its manifestations, is the latest spawn of the technological revolution to capture the popular imagination and transform the way we work, play and live. More than a communications tool, it has become a fashion item, a status symbol and a statement of peer-group membership.

Such is the demand for the accessibility they provide that people are buying them in droves. By December 2000 Australians owned more than 10.1 million digital mobile phones, an 81 per cent leap in ownership in two-and-a-half years. This makes Australians one of the highest users of mobile phones per capita in the world. Industry predicts that by the end of 2001 there will be 315 million mobile-phone subscribers in the world, close to double the number at the end of 1999. Moreover, people are now using their mobile phones more often and at higher power levels, as we move towards the third generation of mobile technology that enables services such as internet connections.

Mobile phones operate using microwave radiation. They receive microwave signals from, and transmit signals to, the nearest phone tower, which is also known as a base station. Similar to radio and television signals, these microwaves travel through the atmosphere and affect people who happen to be in their path. While a mobile phone is turned on, it is sometimes transmitting signals, even when no call is being made, because it keeps in touch with a phone tower. If these signals were visible, we would see ourselves enveloped in a web of radiation enormously dense in high-use areas such as city centres.

Facilitating the mobile system is a network of telecommunications towers or base stations, sprinkled liberally throughout the community. These range from the towering monopoles that loom like pieces of gigantic scaffolding over entire suburbs, to tiny microcell antennas situated on lampposts or shop awnings at just above head-height. Many antennas are strategically placed to avoid notice, hidden in chimneys or church crosses. Some are 'blended' into the environment, disguised as trees or flag poles. All emit microwave radiation.

Base stations are organised in a network of areas called cells. Depending on the user population, these can be anything from tens of metres to 50 kilometres in diameter. Each cell is serviced by a transmitter antenna (often a mobile-phone tower) which sends and receives radio signals at powers of anything from a few watts to 100 watts or more, to or from all mobile phones in its coverage area. As mobile phones move from cell to cell, the signals are automatically transferred from one phone tower to another via a switching centre. This centre also connects signals to the regular telephone network.

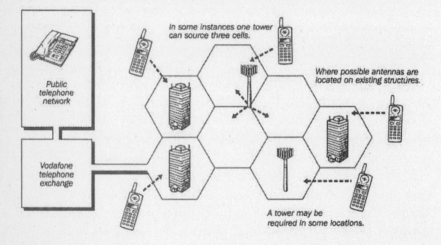

Illustration courtesy of Vodafone

A cellular base station consists of a transmitter/receiver, antennas, an equipment shelter, and cable through which it is connected to the wider telephone system. Some antennas are omnidirectional (transmitting signals in all directions), while others transmit signals in a narrower beam to segments of a cell covering, perhaps, a 120-degree arc. Whereas earlier antennas were arranged in groups of three—one receiver antenna and two transmitter antennas—newer designs house receiver and transmitters in a single panel box.

All signals are transmitted in the radiofrequency band of the spectrum. Digital GSM (Global System for Mobile Communications) base stations operate between 890.2 and 959.8 MHz, whereas the older analogue system operated between 824 and 894 MHz. Third generation (3G) mobile systems operate at higher frequencies of around 1800 MHz.

Because a base station can handle a limited number of calls at a time, when engineers find that capacity is reached, they will divide a cell into two smaller cells, each with its own base station. In particularly dense population areas, such as the inner city or an airport, the volume of usage has necessitated microcells serviced by tiny antennas which can sometimes be seen attached to power poles, traffic lights, awnings, and buildings at just above head height.

A mobile-phone base station emits a signal in a relatively wide beam, rather like a torch beam held almost horizontally to the earth. The intensity of these signals (measured in the centre of the beam) is highest close to the antenna and drops off with distance. On the ground, maximum exposures are generally recorded, not at the base of the tower, but at a distance of 100 to 150 metres away. However, the intensity of the signal will be affected by the contours of the ground and the structures in the path of the beam. A house at the base of the tower may be outside the main beam and may record low readings; a block of flats further away, which is closer to the beam, will record a higher reading.

At present the number of base stations in the Australian community is increasing dramatically. As additional carriers have entered the telecommunications arena, the number of phone antennas in the community is also proliferating.

Does radiation from telecommunications systems pose a health risk?

Electromagnetic radiation from the communications system has not been conclusively proven to cause health problems. Nevertheless, it has been associated with a range of health problems including brain tumours, lymphomas, memory problems, genetic effects, and hormonal changes—which will be discussed further in chapter five.

In addition to the scientific evidence for risk, many people are experiencing inconvenient symptoms or significant health problems from the radiation emitted, particularly by mobile phones.

Is it safe to use mobile phones?

At least 50 per cent of the radiation from a mobile phone is deposited into the user's body. During a call your head will be absorbing just as much radiation as is reaching the base station to connect you with the wider world.

And not only your head is at risk. If you use a handsfree device, you may keep your phone on your belt during a call, with the result that 50 per cent of the signal is being absorbed at the waist. Remember, too, that a mobile phone is transmitting

Yes, our mobile phones are completely safe for the environment! They're made from renewable resources, use rechargeable batteries, and they come in recyclable packaging with a dolphin logo.

sporadic signals to a base station while it is turned on. So it is giving out radiation, even when no call is being made. This means that a phone kept in a pocket or a handbag is irradiating the parts of the body that are closest to it.

The radiation emitted by a mobile phone—and therefore the risk to your health—will vary according to your location. The power at which a mobile phone operates is automatically determined by the nearest base station. When a phone is a long way from a base station it will transmit at higher power, and hence higher levels of radiation will pass through the caller's head, than when when the base station is close. Similarly, a call made inside a building will require more power to penetrate the walls than one made next to an open window or outside the building.

Mobile-phone radiation has been associated with problems such as brain tumours, cancers, lymphomas, altered immunity, effects on DNA, and changes to cells and genes, as you will see in chapters five and six. At present there is no generally accepted scientific proof that mobile phones cause these health problems— just as there was no proof that smoking caused lung cancer for many years. Given that problems such as cancers take many years to develop, proof may still be some time away.

Already there are worrying signs in the community. Many people are reporting a range of consistent symptoms along with more serious problems such as brain tumours. While these reports are a far cry from the scientific proof that's needed, they are an indication that mobile phones may not be as safe as governments and industry persist in telling us; an indication that much more independent research is needed.

Some people are convinced that their tumours are the result of mobile-phone use. In the United States, the stage is set for a rash of litigation against mobile-phone manufacturers. Dr. Christopher Newman, a 41-year-old neurologist with a brain tumour he alleges was caused by the radiation from his mobile phone, is suing a number of telecommunications companies for $800 million. His lawyer, Joanne Suder, has joined forces with Peter Angelos, a highly successful and well-resourced lawyer who litigated successfully against the tobacco and asbestos industries, and who has a team of 110 lawyers in six states across the country. Suder has another ten actions pending. The prospective litigants include a 32-year-old Motorola employee who tested mobile phones each day for over nine years and the widow of a 45-year-old businessman who died with a brain tumour. All three developed tumours adjacent to the position of their phones' antennas.

Several interesting case studies have reported unusual growths in the head that appear to be connected to the use of mobile phones. An English telephone engineer, who was a regular mobile-phone user, developed a rare growth on his salivary gland on the right side of his head where he held his phone. (Pereira and Edwards 2000.) A British dentist reported that a patient developed a 'suspicious lesion' on the right side of his mouth, near the position of his phone, which cleared when he was away from work or used a landline phone. (Watt 2000.)

Not surprisingly, many users are reporting a range of fairly common symptoms from using mobile phones. Several studies have identified symptoms of fatigue, headaches, concentration, and memory problems, warmth around the ear, and tingling around the ear. In a study conducted in Sydney in 2000, I advertised for people who had experienced symptoms while using a mobile phone. I was absolutely deluged with callers. Respondents reported heat on the side of the face used for the call, headaches, pressure, ringing in the ear, dizziness, nausea, and deep pain. Moreover, callers reported that continued use made the symptoms more severe or caused them to appear earlier in the call, so that some of them had stopped using their mobile phone altogether.

In light of this, it may come as no surprise that leading underwriter Lloyd's of London has refused to insure mobile-phone manufacturers against the possibility of legal action in the event that mobile phones do, in the future, become an established health risk. (London *Observer* 11 April 1999.)

It's quite possible that people reporting symptoms from mobile-phone use are the canaries in the communications coalmine, sounding an early alarm. Their unified voice and the scientific evidence of risk suggests it may not be a bad idea to reduce your exposure—and some suggestions for doing so are outlined in chapter eight.

Is it safe to use mobile phones while driving?

The metal shell of a car—or, for that matter, a train, bus, taxi, or lift—reflects the signal of a mobile phone so that it bounces around the vehicle, irradiating not only the user but also other passengers. In a car, you can at least reduce your exposure by using a handsfree kit with an external aerial. However, it is strongly advisable not to install an antenna near the rear window of a car, particularly near a baby seat.

Quite a number of studies have now shown that using a mobile phone while driving, whether or not it is handsfree, increases the likelihood of an accident. There is also evidence that the accident rate increases, not just during a mobile-phone conversation, but for a short time afterwards.

In 1997 Japan's National Police Agency reported that the use of mobile phones while driving had led to an increased rate of traffic accidents. Its figures showed that drivers using mobile phones caused 2,297 road accidents that killed 25 and injured 3,328 people. (*Sydney Morning Herald* 20 March 1998.)

In the UK, an independent expert group on mobile phones, chaired by Sir William Stewart, concluded from an examination of available research that conversing on a mobile phone 'impairs driving performance' and increases the risk of an accident. It therefore concluded that 'the detrimental effects of hands-free operation are sufficiently large that drivers should be dissuaded from using either hand-held or hands-free phones whilst on the move.' (Recommendation 1.22.)

The World Health Organisation has also recommended that mobile phones not be used while driving. In its Fact Sheet no.193 of June 2000, it states, 'In moving vehicles there is a well established increase in the risk of traffic accidents while the driver is using a mobile phone, either a conventional handset or one fitted with a 'hands free' device. Motorists should be strongly discouraged from using mobile phones while driving.' (http://www.who.int/inf-fs/en/fact193.html.)

Of course, Harriet dear, you have my undivided attention. Now what were you saying about the wedding guest list?

Studies that have found mobile phones to be a risk when driving can be found in Appendix A on pages 181–2.

Should children use mobile phones?

Children are often regarded as particularly vulnerable to the radiation from mobile phones. One reason for this is that children have thinner skulls than adults, and radiation is able to penetrate further into their heads. As children and foetuses grow, their cells are often in a process of dividing (called mitosis), and this is the very stage at which they are most vulnerable to radiation. Children may also be affected by the radiation from a mobile phone because their heads are about the size of the waves generated by the phones; this can result in an interaction known as a resonance effect (see chapter six). Some mobile phones emit signals that lie in the range of alpha and delta brain waves—the very brain patterns that are constantly changing in children until the age of about twelve, when the alpha rhythm becomes established. Finally, unlike their parents, children of the present generation have the possibility of a lifetime of exposure and the accumulation of many decades of effects, a situation that has no historical precedent.

Yet, ironically, as competition for sales escalates, it is to the lucrative market of children that manufacturers are turning their commercial attention. Just before Christmas 1999, a large toy-retailer distributed a leaflet promoting the use of mobile phones by children. The advertisement used kids-speak to promote 'xtreme mobile phone deals' which included 'no age limit' and prepaid charge cards, and featured two 'cool' teenagers using mobile phones.

By coincidence, the advertisement appeared in the same week as details about Dr. Om Gandhi's study which showed that children absorb up to 50 per cent more radiation than adults because of their thinner skulls. (Gandhi et al. 1996.)

Even toddlers have been considered fair game in the corporate battle for potential customers. A visit to the same company's web site revealed a range of toy mobile phones specifically designed for pre-schoolers. There was the Talking cellphone with 'flip-down mouth piece and clicking antenna'. The Elmo cell phone allowed children to 'pretend to make or receive important calls. The phone features flashing lights, fun phone sounds, and five Elmo phrases!' Traditional icon, Barbie, had been updated with a 'cool electronic keychain [that] features a cell phone with authentic sound effects, music, lights, and the voice of Barbie!' Finally, the Tech-Link cell 200 walkie talkies were 'styled just like trendy flip phones!' (*EMRAA News* December 1999.)

Such advertising creates the impression that mobile phones are socially desirable and safe.

Clearly, while the safety of mobile phones is in serious doubt, the promotion of their use by children is irresponsible to the point of criminal recklessness.

In Britain a report on the risks of mobile phones by Sir William Stewart in April 2000 recommended that children be discouraged from their use. '… children may be more vulnerable because of their developing nervous system, the greater absorption of energy in the tissues of the head, and a longer lifetime of exposure. In line with our precautionary approach, at this time, we believe that the widespread use of mobile phones by children for non-essential calls should be discouraged. We also recommend that the mobile-phone industry should refrain from promoting the use of mobile phones by children.' (Recommendation 1.53.)

Similarly, the report by the chair of the Australian Senate Inquiry into EMR in 2001 stated that 'the effects of RF radiation on children should be treated as a priority research area given the increasing use of mobile phones by children and teenagers.' (*Inquiry into Electromagnetic Radiation* 2001.)

Do mobile phones interfere with electronic equipment?

That mobile phones interfere with delicate electronic equipment is common knowledge. After all, we are accustomed to being asked to turn off the mobile phone or laptop on an aircraft, or seeing signs at the post office warning phone users that 'Mobile phone signals interfere with our computers. Please switch them off whilst in this office.'

You can observe how the radiation from a mobile phone interferes with electronic circuitry in a simple experiment. Turn your mobile phone on and leave it next to your radio for a few hours. Whenever you hear static from your radio, you know the mobile phone is cutting in and interfering with the signal. In much the same way, the radiation from a mobile phone interferes with the electronic circuitry of other devices—and perhaps with that of our brain.

There is evidence that mobile phones interfere with the navigational systems of aircraft; indeed, all Australian airlines presently direct passengers not to use their mobile phones during a flight. The Civil Aviation and Safety Authority is presently considering changes to regulations that would impose permanent bans on mobile-phone use on aircraft to bring Australia into line with international regulations.

Electromagnetic interference from mobile phones is a particular problem for people with pacemakers and cardiac defibrillators. Many studies have shown that mobile phones—both analogue and digital—and walkie talkies cause interference with the devices, and have recommended that wearers keep the devices away from the heart or avoid them altogether. A list of these can be found on pages 183–4. Similarly, pacemakers and defibrillators have been affected by the radiation-gates now appearing in ever-increasing numbers in department stores and supermarkets. As medicine is increasingly implanting electrical devices in patients, the issue of electromagnetic compatibility is likely to become an increasing problem in the future.

Mobile-phone base stations

Providing the connection for Australia's ten million mobile phones is a vast array of mobile-phone base stations dotted throughout the cities. At present there are estimated to be approximately ten thousand mobile-phone base stations in Australia, and the number is growing rapidly as the newer players to the market establish their own networks and third-generation technology demands more, though smaller, antennas.

Low-impact base stations

The Federal Government has established two arbitrary categories of base station. The first is the 'Low Impact' category, classified under the 1997 Low-Impact

Determination, comprising antennas which are generally less than five metres in height. This classification, based misleadingly on appearance only, has nothing to do with the amount of radiation the antennas emit or the risk to public health. Classifying a radiation-emitting structure according to its appearance and not its emissions is rather like classifying a pharmaceutical drug according to the colour of its packaging.

Go ahead, Adam. It's safe to eat. It's under five metres long. It's just a low impact apple.

After 1997, legislation allowed telecommunications carriers to erect so-called 'low impact' base stations with immunity from state and council regulations. There was no requirement to notify councils or even affected communities.

By providing carriers with a carte blanche to erect 'low impact' antennas virtually at will, the government created a monster to confront and confound the community. In an effort to establish a competitive network in haste, common courtesies were often sacrificed to the great god of corporate profit. Tenants found first one, then another antenna erected on their roof without so much as a word of explanation. Entire communities gathered to voice their opposition to a proposal, only to be told by a carrier that the proposal would proceed regardless. New antennas were even erected in positions that exposed maintenance workers of other carriers to danger on site.

Still antennas proliferate on tall buildings of every description, in neighbours' yards, on lamp-posts outside people's bedrooms and next to schools. And communities are powerless to stop them.

In the pictures opposite, two 'low impact' antennas were installed on top of a block of flats and unsightly cables run up the side of the building. Residents—a family of three and a family of four—received no notification whatsoever. Near the building is a 'not low impact' tower, and all of these facilities are within a short distance of the local primary school of 400 children.

Needless to say, there has been a strong community backlash to legislation that allows 'low impact' antennas to be installed without council or community approval. In Brisbane in mid-2000, the early stages of escalating tower construction engendered a vigorous response. Within five weeks, no less than 16 community groups were established to oppose the proliferation of mobile-phone base stations and to demand legislative changes that would establish 800-metre buffer zones from sensitive areas. In New South Wales (NSW), Andrew Tink MP expressed his concerns in state parliament: 'I am convinced that these so-called low-impact facilities are nothing of the sort when we are talking about their electromagnetic impact. I believe strongly that they should be the subject of some form of community consultation and local consent process. I place on the record that I am lobbying the Federal government to try to achieve further developments in this area.' (*Hansard* 14 April 2000.)

In August 2000, federal shadow communications minister Stephen Smith demanded legislative changes that would protect the community, and his call was endorsed by the Australian Local Government Association (ALGA). 'Local Government and communities across Australia are currently disempowered under existing telecommunications legislation [Low Impact Determination],' said ALGA President Cr. John Ross. 'Although some telecommunications carriers consider potential mobile-phone tower installations in consultation with local authorities and affected communities, there are some maverick newcomers to the industry that continue to abuse current legislation.

'Local Governments nation-wide have contacted ALGA with their concerns that some carriers are riding roughshod over communities when deploying their networks.' (Press release 26 July 2000.)

The 2001 chair's report on the Senate Inquiry into EMR recommended that 'the Government review the Telecommunications (Low-Impact Facilities) Determination 1997, and as a precautionary measure, amend it to enable community groups to have greater input into the siting of antenna towers and require their installation to go through normal local government planning processes.' (Recommendation 2.5.)

Not-low-impact base stations

Base stations that do not fall into the above category are known by the tautological description of 'not low impact facilities'. These do, at present, require council approval to be erected. Needless to say, councils have been under considerable pressure to reject applications regarded by the community as being inappropriate or dangerous.

In an effort to address the community's concerns, some councils developed policies on the siting of mobile-phone base stations. The first policy, devised by Dr. Garry Smith of Sutherland Shire Council in 1997, provided that base stations should not be erected within 300 metres of residences, schools or other sensitive areas, unless average annual exposure could be kept under 0.2 microwatts/cm^2 (μW/cm^2). Though this policy did not have legislative force, carriers demonstrated that they were willing to comply with it. Based on its success, many councils throughout the country adopted broadly similar policies.

However, the proliferation of council policies created difficulties for carriers who were obliged to comply with different requirements in different localities. The federal government responded by requiring each state government to develop a single policy on the siting of mobile-phone base stations that would supplant those of local councils. There are concerns that, when fully implemented, this will effectively disempower local councils and local communities from having a say about the location of radiation-emitting infrastructure in their own domain.

In early 2001, many of the state policies were still in the process of being developed. So far Tasmania has a policy that is incorporated into state legislation. In Victoria, a consultant developed a code of practice that became effective on 1 July 1999 and that does not fully take into account the issue of health risks from EMR. In NSW, the same consultant is presently drafting a policy that will most likely be introduced in late 2001.

Also during 2000–1, a Code of Practice on the Siting of Telecommunications Infrastructure was drafted. At the time of writing, this code requires carriers to notify councils and communities about any base stations they intend to install, and to engage in a process of community consultation. It requires carriers to consider environmental issues and sensitive areas, and to operate their facilities at the lowest possible power. However, neither the code nor legislation empower councils or communities to reject applications for 'low impact' facilities, which has led critics of the code to question the authenticity of the 'consultation' and the precautionary claims of the document.

Where are mobile-phone base stations located?

Mobile-phone antennas can been seen as free-standing monopoles or adorning the roofs of flats, offices, hospitals, and tall structures throughout our cities. While many of them are easy to locate, some have been skillfully 'blended' into the environment, disguised as pine trees, flag poles, and church crosses, so that they evade detection by even the most skillful scrutineer. While hiding antennas in this way may reduce people's concerns about their visual impact, it does nothing to allay their health concerns. In fact, many people wish to know where these facilities are located in relation to their homes or to a property they are considering buying.

It is possible to access information about the location of all mobile-phone base stations and antennas in Australia. The Australian Communications Authority has established a web site with this information which is updated every three months. It can be found at: http://www.aca.gov.au/database/radcomm/index.htm.

Is it safe to live near a mobile-phone base station?

Even if the sort of radiation emitted by mobile phones and mobile-phone base stations is dangerous, could it be a risk at the very low levels emitted by an antenna?

While the government and carriers assure us that there is no risk to health, that is not necessarily the case.

In mid-1999, Dr. Leif Salford stated publicly that exposure to the radiation from mobile phones at levels you would receive near a base station can cause the protein albumin to leak through the blood-brain barrier, and suggested this might be responsible for neurodegenerative diseases such as Alzheimer's. According to Salford, 'Radiation from the base stations of the mobile-phone systems should therefore be enough to affect the brain. Even someone who is in the vicinity of a person making a call may be influenced by the radiation from the phone.' (*Svenska Dagbladet* 16 September 2000.)

Even though no studies have been conducted on populations living near mobile-phone base stations, there is some evidence that the low levels of radiation they emit is harmful to health—and can even affect plants and animals. The following is a sample of those studies which have found effects at levels hundreds of times below those allowed by the standard:

Prausnitz, 1962	Mice exposed to 9.27 GHz had testicular damage at a mean of $0.22 \ \mu W/cm^2$.	Prausnitz, S. and Susskind, C., 'Effects of Chronic Microwave Irradiation on Mice', *IRE Trans on Biomed. Electron.*, vol. 9, pp. 104–8, 1962.
Schwartz, 1990	Cells exposed to EMR at 240 MHz showed calcium ion efflux at $0.08 \ \mu W/cm^2$.	Schwartz, J.L. et al., 'Exposure of Frog Hearts to CW or Amplitude-modulated VHF Fields: Selective Efflux of Calcium Ions at 16 Hz', *Bioelectromagnetics,* vol. 11, no. 4, pp. 349–58, 1990.
Goldsmith, 1995	Embassy workers exposed to EMR developed neurological symptoms, abnormal blood cell	Goldsmith, J.R., 'Epidemiologic Evidence of Radiofrequency Radiation (Microwave) Effects

	counts, chromosome aberrations and cancer at 1–2 µW/cm².	on Health in Military, Broadcasting, and Occupational Studies', *Int. J. Occup. Environ. Health,* vol. 1, pp. 47–57, 1995.
Alpeter, 1995	People living near a shortwave transmitter experienced sleeping problems, nervousness, joint pain, and congestion at exposures of 0.04–3.8 µW/cm².	Alpeter, E.S. et al., *Study of Health Effects of Shortwave Transmitter Station of Schwarzenburg, Berne, Switzerland,* University of Berne, Institute for Social and Preventative Medicine, August, 1995.
Maes, 1996	Blood samples exposed to a carcinogen and EMR at levels experienced near a mobile phone tower showed evidence of chemical mutation.	Maes, A. et al., '954 MHz Microwaves Enhance the Mutagenic Properties of Mitomycin C', *Environ. Mol. Mutagen,* vol. 28, no. 1, pp. 26–30, 1996.
Hocking, 1996	People living near TV transmitters in Sydney had a higher rate of leukemias. Outdoor exposure was 0.2–8 µW/cm², and exposure was less indoors.	Hocking, B. et al., 'Cancer Incidence and Mortality and Proximity to TV Towers', *Med. J. Aust.,* vol. 165, nos. 11–12, pp. 601–5, Dec., 1996.
Kolodynski, 1996	Children living closest to the Skrunda transmitter had 'less developed memory and attention, slower reaction times, and decreased endurance of neuromuscular apparatus' at exposures of 0.0008–0.41 µW/cm².	Kolodynski, A.A. and Kolodynska, V.V., 'Motor and Psychological Functions of School Children Living in the Area of the Skrunda Radio Location Station in Latvia', *Sci. Total Environ.,* vol. 180, no. pp. 87–93, 1996.
Balodis, 1996	Pine trees near the Skrunda radar showed reduced radial growth at 0.011–0.41 µW/cm².	Balodis, V. et al., 'Does the Skrunda Radio Location Station Diminish the Radial Growth of Pine Trees', *Sci. Total Environ.,* vol. 180, no. 1, pp. 57–64, 1996.
Balode, 1996	Cows near the Skrunda radar station showed chromosome	Balode, Z., 'Assessment of Radio-frequency

	and reproductive changes at 0.042–6.6 µW/cm^2.	Electromagnetic Radiation by the Micronucleus Test in Bovine Peripheral Erythrocytes', *Sci. Total Environ.*, vol. 180, no. 1, pp. 81–5, 1996.
Persson, 1997	Blood brain barrier in rats became permeated when exposed to EMR at the levels that would be experienced near a base station.	Persson, B.R.R. et al., 'Blood-Brain Barrier Permeability in Rats Exposed to Electromagnetic Fields Used in Wireless Communication', *Wireless Network*, vol. 3, pp. 455–61, 1997.
Kwee, 1997	Cells exposed to EMR at SARs of 0.000021–0.0021 W/kg (the level a person would receive at 30–4960 m from a base station) showed changes in proliferation.	Kwee, S. and Rasmark, P. 'Radiofrequency Electromagnetic Fields and Cell Proliferation', Second World Congress for Electricity and Magnetism in Biology and Medicine, Bologne, Italy, 1997.

But what of other serious problems such as cancers and leukemias? Mobile telecommunications systems are comparatively recent developments—just over a decade old—and cancers and leukemias can take many decades to develop. If there is a connection, it is not likely that the ramifications will be experienced for many years to come.

However, already there are worrying signs.

Although anecdotal evidence leaves a lot to be desired, it is at present the only indicator of what is actually happening to people living or working near mobile-phone base stations, as absolutely no research is being done on this vitally important issue.

Over the last three years, I have had calls from many people claiming that they—and sometimes also their neighbours—have developed serious health problems since the installation of a mobile-phone base station. In one neighbourhood, three women within metres of a tall mobile-phone base station developed breast cancer.

One caller described an unusual 'cluster' of cancers that developed underneath a mobile-phone base station. He reported that in one office of 20 workers situated directly underneath an antenna, two workers had developed breast cancers and two developed other cancers after the antenna was installed.

Similarly, three workers in a suburban supermarket developed throat cancers around the same time as each other. According to the doctor consulted, an environmental factor was likely to be implicated. One environmental factor which caught the attention of my informant was the forest of antennas on the roof of the building where the employees worked.

Cancer is not the only problem reported. Just after a mobile-phone base station was installed 70 metres from her home, a young epileptic who had previously experienced only one or two fits a month began to have severe fits four to six times a day. Coincidence? Probably not, because the number of fits that the girl experienced dropped dramatically when she was away from home and when the transmitter was turned off.

One call which left an indelible impression on me was from a man in Queensland who reported that, en route to a holiday in Cairns, his wife experienced excruciating 'seizure like' symptoms every time the car passed a mobile-phone base station. His dilemma was how to get her home again—alive.

A number of callers have reported unpleasant sensations from the radiation from a mobile-phone base station. These range from heat and head pressure to nausea and dizziness. One caller described how being anywhere near a base station affected her so badly that she was obliged to travel some distance out of her way to reach the local town to avoid passing by an antenna. Once a bus in which she was a passenger parked near a base station for ten minutes. My caller reported such an intense reaction that she was subsequently bed-ridden for a week.

A Sydney community sandwiched between two base stations (one of which comprises an array of antennas, and the other accommodates four carriers) has developed an assortment of symptoms. Two residents have recently died of cancer. A third experiences a range of health problems consistent with radiation exposure, including memory problems, sleep problems, panic attacks, and low immunity. A fourth resident has microwave hearing (a well-known phenomenon in which people report sounds from transmitted signals). Finally, two trees growing directly underneath one of the antennas developed a large amount of dead foliage on top, but not underneath.

Even animals are exhibiting effects. Dr. Wolfgang Loscher reported an unusual range of symptoms of a herd of dairy cows pastured near a television and mobile-phone transmitter. In addition to exhibiting rather strange behaviour, the cows produced less milk than normal and experienced more health problems. One cow, which had behaved abnormally in her regular pasture, reverted to completely normal behaviour within five days of being taken to a stable 20 kilomtres away. When returned to the field near the antennas, her symptoms returned. Loscher was unable to identify any factors to account for the strange symptoms other than the presence of the antennas, and concluded that it was possible that the electromagnetic fields they emitted were related to the problems the cows were experiencing. (The cows would have been receiving exposures of 0.02–7 mW/m2.) (Loscher and Kas 1998.)

Reports such as these are generally dismissed as anecdotal—the implication being that they are unworthy of notice. However, given that neither the government nor the carriers have ever made a serious effort to monitor or to study the health of people (or animals) living near mobile-phone base stations, they are not in a strong position to deny the existence of such effects.

One of the factors that may affect the safety of people living or working near a mobile-phone tower is the existence of hot spots. These can be created by the overlapping of signals from various facilities or by features of the environment itself. Metal structures such as doorframes, window frames, metal beams in buildings, cars, and children's play equipment can all amplify the signal from a base station. Sydney University's Professor David McKenzie has taken measurements in a new flat built directly opposite an existing base station. He found that the fields in the lounge room were 1.1 V/m, but near the metal-framed window they were 2.7 V/m. On the outside verandah they were 6.6 V/m, but near the metal doorframe they were 17 V/m. (RF Spectrum Conference 2001.) The amplification effect needs to be taken into consideration when assessing proposals for new base stations.

Finally, it needs to be pointed out that mobile-phone base stations have an almost rabbit-like tendency to multiply. Because the government encourages carriers to co-locate (establish new antennas on an existing site), the construction of a new tower will more than likely attract other carriers with their transmitters. This means, of course, that emission levels will increase, often exponentially.

Are some people more vulnerable?

Like food, music or chemicals, EMR affects people differently. Generally speaking, children, foetuses, and sensitive people are more vulnerable than young healthy members of the community, as are those in highly exposed occupations.

As previously mentioned, children are more susceptible to the effects of radiation than adults because their rapidly dividing cells are more vulnerable, they have thinner skulls, and they absorb more radiation than adults do. People with cancer are more at risk because cancers have unusual electrical properties and absorb more energy than normal tissue. Some of the more sensitive members of the public already report strong reactions to exposure from mobile-phones or base stations, as we have seen, and this reaction is likely to grow as the technology proliferates.

What are the risks to my property value?

There is evidence that, like high-voltage powerlines, mobile-phone towers devalue adjacent properties:

* A jury in the 295th Judicial District Court of Harris County Texas awarded $1.2 million to a couple who claimed that a 100-foot mobile-phone tower invaded their privacy and created a nuisance. The damages payment included compensation for loss of use and enjoyment of the couple's property, mental anguish, and legal fees. (*Houston Business Journal* 25 February 1999.)

* A study at Lookout Mountain, Colorado in the US found that property values declined associated with the perceived medical harm from a proposed high-definition television broadcast antenna. The author wrote: 'The perception of reality is that living within a five mile radius and having a direct line of sight of a high-definition television antenna can cause medical harm to human beings. This perception may not be based on medical or scientific fact but causes concerns within the community. This perception of reality makes property within the effected zones less attractive to both existing occupants as well as prospective new buyers.' (Hutchison 1999.)

* In Ireland, a valuation of properties nearest a mobile-phone mast found that 'they had fallen by around 25 per cent'. (Belfast Telegraph Newspapers Ltd. 15 January 1999.)

* In Germany, the monthly rent of an apartment was reduced by 20 per cent because mobile-phone antennas had been installed on its roof without notification and agreement of the tenant. (*Amtsgericht München* 27 March 1998.)

In Australia

* A beautiful home in Queensland located close to a tall mobile-phone base station took three years to sell and finally sold for much less than the original market value.

* At the Senate Inquiry into Telecommunications Legislation in 1997, the three carriers then in operation—Telstra, Optus, and Vodafone—all admitted that mobile phones made a 'small' impact on property values. (*Hansard* 14 February 1997.)

What levels of exposure are safe for mobile phones and base stations?

The RF standard

In Australia, the Australian Communications Authority (ACA) requires industry to comply with the exposure limits of the now defunct Australian Standard

(AS2772.1) which allows the public to be exposed to 200 µW/cm^2 from mobile-phone antennas and 417 µW/cm^2 from mobile phones. At the time of writing, a working group of the Department of Health is developing a new standard based on the limits of the Guidelines of the International Commission of Non Ionising Radiation Protection (ICNIRP). These limits are more relaxed than previous standards, and allow people to be exposed to over twice as much radiation from GSM mobile phones, and four-and-a-half times as much radiation from higher-frequency 3G phones.

These standards, like their predecessors and most overseas contemporaries, are thermal ones. That is, they are based on the premise that health problems only occur if the body heats up by one degree Celsius. While they recognise that bio-logical effects have been shown to occur at lower (athermal/below heating) expo-sures, they do not believe these biological effects to be necessarily significant for health.

Yet some of the effects that have been observed at athermal levels of exposure include effects on the brain, behaviour, performance, learning, sleep, reproduction, cancer, DNA, immunity, hormones, cell proliferation, and genes. It is hard to see how some of these problems could not be significant for health.

Veteran EMR researcher Dr. Ross Adey believes that we now have substantial evidence that athermal levels of EMR can be a risk. 'The laboratory evidence for athermal effects of both ELF and RF/microwave fields now constitutes a major body of scientific literature in peer-reviewed journals. It is my personal view that to continue to ignore this work in the course of standard setting is irresponsible to the point of being a public scandal.' (Adey 1995.)

Even more evidence has become available since Adey wrote this statement in 1995.

Another extremely significant problem with the RF standards is the assump-tions they make about how much radiation the body can absorb. Both standards are based on the assumption that a person's body can absorb four watts of radiation per kilogram (4 W/kg) for about 30 minutes before its average temperature rises by 1° Celsius, causing health problems. Now it is quite possible to expose certain parts of a person's body to 25 times as much radiation (up to 100 W/kg) without the overall temperature of the body rising above the 1° Celsius mark. Therefore, the standards assume that it is quite safe to do this, and that parts of the body can be exposed to very much more radiation than the whole body (as occurs, for example, with mobile phones).

However, these assumptions leave some very important questions unan-swered. For example:

- can children absorb 25 times as much radiation as adults to the head?
- do all people respond in the same way, or do we absorb radiation differently?
- can pregnant women dissipate heat as effectively as other people?
- do we dissipate heat as well in summer or after exercise?
- how do we know that adverse health effects don't occur if only parts of the body (particularly the brain) are heated to 100 W/kg?

Allowing parts of the body, such as the brain, to be exposed to very high levels of radiation may not altogether be a good idea. How do we know that there will not be localised damage? If, for example, we put a finger in a candle flame, the body will not heat up by 1° C, but there will be some quite obvious localised damage. The finger will burn and blister, reminding us to be more careful next time.

Dr. Henry Lai has found that different parts of the brain do respond differently to radiation. He exposed rats' brains to pulsed radiation at 2450 MHz for 45 minutes to examine the effects on some neurotransmitters. He found that in two parts of the brain (the hippocampus and frontal cortex) there was a decrease in take up of high-affinity choline, but there was no significant change in choline uptake in three other areas (the striatum, hypothalamus, and inferior colliculus). (Lai et al. 1997.)

Both the existing and the proposed Australian standards incorporate 'safety factors' which they claim provide protection for workers and additional protection for the general public. For workers, there is a safety factor of 10, which limits their exposure to 0.4 W/kg over a 40-hour week.

For the general public, there is a safety factor of 50, which limits their exposure to 0.04 W/kg. However, because members of the public are exposed to radiation for every one of the 168 hours of each week, they are exposed for almost five times as long as workers. This negates any so called additional safety factor.

Not only are there theoretical problems with the standards for RF exposure, but there are experiential problems. Many people are reporting symptoms from exposure to the radiation from mobile phones—and even mobile-phone towers—and this strongly suggests that the present standards are not providing adequate protection.

These issues raise the question of whether it is appropriate to wait until science has a full understanding of exactly how EMR affects the body before we take precautions. Or do we take precautions now based on the evidence of risk and the experiences of the community?

What do you mean the community doesn't think the standards committee is representative? It has Optus, Telstra, Telstra NZ, Telecom NZ, Vodafone, the Telecommunications Users' Group, the Calathumpian Mobile-Phone Network, the Federal Ministry for Profit ...

Alternative views

If the existing standards for RF exposure are inadequate, what is the alternative? Unlike the situation with powerlines, it is not clear just what level of exposure to microwaves represents a risk.

New Zealand biophysicist Dr. Neil Cherry, who has analysed the work of over 600 researchers in this field, has concluded that the level of allowable microwave exposure needs to be enormously reduced. According to his investigations, the threshold level of exposure is 0.06 $\mu W/cm^2$ for cancer and 0.0004 $\mu W/cm^2$ 'for sleep disruption, learning impairment and immune systems suppression', which are less than the levels Australians and New Zealanders are already experiencing. Based on this, he advocates setting an exposure limit of 0.05 $\mu W/cm2$ to be reduced to 0.01 $\mu W/cm^2$ in ten years. (Cherry 1999.)

In June 2000 a number of experienced EMR researchers met in Salzburg to consider the issue of mobile-phone antennas. While they could identify no 'threshold for adverse health effects', they nevertheless suggested some precautionary limits. 'For the total of all high frequency irradiation a limit value of [10 $\mu W/cm2$] is recommended. For preventive public health protection a preliminary guideline level for the sum total of all emmissions from ELF pulse-modulated high-frequency exposure facilities such as GSM base stations of [0.1 $\mu W/cm2$] is recommended' (see chapter seven).

The general lack of agreement on a 'safe' level has made it extremely difficult to legislate to protect people. Previously, council policies were a workable compromise which afforded some protection for the general public, while allowing carriers to erect their networks.

Social effects

In a society where time has become a precious—and scarce—commodity, the mobile phone is the ultimate time-saver. Because you can use it to access the internet, to clinch that business deal between courses, and to mollify clients after sundown, it allows you to work from home, the car, the train, the pavement—for as many hours a day as you—or your employer—wish.

But what are the social ramifications of this technology?

A study by social researcher Brian Sweeney has found that many Australians feel overwhelmed and overburdened by technological developments, and suffer from what he calls 'digital depression'. He also found that working hours have increased, with the average person now working a 50-hour week. 'Consumers feel they are disappearing into the deep, all-encompassing oceans of globalisation, technology, information and big business,' he writes. 'Their lives are out of balance, they are time poor and stress rich.' (*Sun-Herald*, 23 April 2000.)

Mobile phones have become the social epidemic of the new millennium.

As I was saying, George, if you get me a mobile phone, I'll be able to keep in touch with you the whole time.

Imagine that from every mobile phone in the world that is presently turned on, whether or not a call is being made, there is a red cord reaching to a mobile-phone base station. Imagine red cords connecting one base station to another across vast distances. Imagine red cords from some of the newer computers accessing the internet or sending an email to another computer, perhaps on a different continent. Imagine a red cord from every television and radio tower to every home in that city, right across the world. Imagine millions of interwoven red cords stretching across every continent, permeating the atmosphere, penetrating people, enormously profuse in cities. Imagine the web of cords growing denser with each passing month.

Now imagine that each of these cords has energy that interacts with everything in its path—people, plants, animals, chemicals, and other sources of energy.

You begin to have a sense of the magnitude of this technology, of its pervasiveness, of its risks. In a single generation, we have created a web of artificial energy that has saturated the entire planet, has touched every living creature on it, and has created an electro-chemical cocktail of unknown consequences.

Tragically, this invisible web is also screening us from the real effects of electromagnetic radiation. Because every person on this planet is swimming in a sea of radiation, there are literally no suitable controls available for scientific experiments. There is no way of comparing exposed populations to unexposed populations, simply because there are no unexposed, unaffected populations available anywhere. This is what makes health effects so difficult to track with existing statistical methods. And this is what makes governments' and industries' denial of health effects so blatently inappropriate.

Although there may be little you can do to protect yourself from the radiation coming from a mobile-phone base station, there are precautions that can be applied to the establishment of infrastructure, and these will be discussed further in chapter seven. For precautions you can adopt to reduce your exposure to the radiation from mobile phones, see chapter eight.

The great health debate

-5-

Does EMR cause health problems?

Having a problem with your health is not just a question of being covered in blotchy pink rashes, being in the grip of a raging fever or in the last stages of a terminal disease. Being unwell includes being rundown, being under par, being anything less than your full potential. The world 'health' derives from the Old English 'haelth' which relates to the world 'hal' or 'whole'. So historically, being healthy meant being whole.

The connection between health and 'wholth' has some official status. The 2001 Chair's report of the Senate Inquiry into EMR adopted the definition of health used by the Stewart Inquiry, which in turn adopted it from the World Health Organisation. All three have regarded health as 'the state of complete physical, mental and social well-being, and not merely the absence of disease or infirmity'.

The important question is, does electromagnetic radiation compromise our 'wholth'? Does exposure to appliances, powerlines, computers, mobile phones, and mobile-phone towers adversely affect our well-being and does it contribute to disease?

A large body of scientific evidence indicates that a range of symptoms—from headaches and insomnia to brain tumours and leukemias—is associated with exposure to EMR. At present, most research has concentrated on more serious problems such as cancers and leukemias, which can take up to forty years to develop. However, there is evidence that the effects extend to a wide range of functions associated with many systems of the body. There is also evidence that EMR contributes to symptoms of headache, fatigue, fuzzy-headedness, memory problems, and depression, among others.

While a great many studies show that EMR does affect the body, there are many that show no effects, and some that show beneficial effects. This apparent contradiction is the basis on which industry and governments justify their claims that there is 'no conclusive proof' of adverse effects. Sometimes they have gone so far as to say there is *no* evidence of health effects.

What health effects?

However, there are good reasons why studies have not always found an association between EMR and health problems, as you will read in the next chapter.

In this chapter we look at some of that 'non existent' evidence: at some of the many studies that have found an association between exposure to EMR from powerlines or radiofrequency sources and a range of health problems. Many more studies are listed in Appendix A. The purpose of these summaries is not to suggest that any individual study *proves* that EMR causes health problems—with the current state of knowledge it is almost impossible to prove a causal connection— but to highlight the trends that have become apparent in the scientific literature, and to show that there are very good reasons why we should take sensible precautions to reduce our exposure.

Brain tumours

At least 66 studies have found evidence that there could be an increased risk of brain tumours as a result of exposure to EMR across the non-ionising part of the spectrum. Many of these have shown that people occupationally exposed to EMR from power sources or microwaves have an elevated risk of such tumours while others have shown that living in high fields from powerlines increases the risk.

Several studies have concentrated on workers exposed to high levels of EMR from the power system. Dr. R. Lin found that electrical workers had 2.8 times the usual rate of brain cancers, and that highly exposed workers died younger. (Lin et al. 1985.) David Savitz found that workers in an electrical utility had an increased rate of the cancers, and that the risk increased with the duration of employment and the strength of the magnetic field. (Savitz et al. 1988.) Susan Preston-Martin found that exposed workers had over four times the rate of brain tumours. (Preston-Martin et al. 1989.) Birgitta Floderus found an increased risk for workers, and that the risk increased with exposure. (Floderus et al. 1993.)

It is not just the magnetic field that has been associated with brain tumours. A study by P. Guenel found that workers in an electrical utility exposed to electrical fields of 13 V/m for 25 years or more had a seven-fold increase in the disease. (Guenel et al. 1996.)

Brain tumours have also been associated with exposure to EMR from radiofrequency sources. Dr. Szmigielski found that Polish military personnel exposed to this type of radiation had more than double the average rate of brain cancer. (Szmigielski 1996.) Beall found that people who used VDTs for more than ten years had an increase in brain tumours (Beall et al. 1996), and P. Ryan found a similar increase for people who worked with cathode-ray tubes. (Ryan et al. 1992.)

But does the use of mobile phones increase the risk of brain tumours?

When, in 1993 David Reynard commenced legal action in the US after his wife died of a brain tumour he alleged was caused by her mobile phone, shock waves rippled throughout the mobile-phone industry, company share prices plunged, and the issue became firmly etched in public consciousness. Since then many other people have claimed that their tumours were caused by mobile-phone use, and some have taken legal action. Though most of the cases to date have been withdrawn or settled out of court, lawyers in the US are entering a new round of legal activity (see chapter four).

In early-2001 three studies on the incidence of brain tumours among mobile-phone users were released which were promoted worldwide in the media as exonerating the devices. 'Mobiles Get Clean Bill of Health,' read a headline in *The Australian*. 'Mobile Phones Don't Cause Brain Cancer or Leukemia, Study Finds,' said a report in Bloomberg.' A Danish study led by Christoffer Johansen studied 420,000 Danes who had used mobile phones for up to 18 years, and concluded they had no increased risk of cancer. However, the study, financed by Danish telecommunications companies, had several design flaws and did not conform to the research protocols recommended by the World Health Organisation. (Johansen 2001.)

Two US studies also showed 'no risk'. Dr. Joshua Muscat found that, after interviewing 469 subjects with brain cancer, and controls, there was no difference in mobile-phone usage. Dr. Muscat, J. Peter Inskip, and Martha Linet also found that there was no real difference in mobile-phone usage among subjects who had a brain tumour and those who did not. (Inskip et al. 2001.) Both of these studies considered only the short-term effects of mobile-phone use and, as we have seen, cancers can take many years to develop. According to Dr. George Carlo, the studies did not even consider the right kind of brain tumours. 'Tumors in almost all patients were located in interior regions of the skull that couldn't be reached by cellphone radiation, which penetrates only 50 millimetres inside adult skulls. In other words: The statistical studies only proved that tumors that couldn't be reached by cellphone radiation weren't caused by cellphone usage. That's hardly reassuring.' (letter, 3 April 2001.)

Studies such as these tend to deflect the concerns of the community without adding substantially to our understanding of the issue.

In mid-2001, two Swedish researchers announced that they had found a higher rate of brain tumours among the users of older-style analogue mobile phones. Dr. Lennart Hardell and Dr. Kjell Johannsen Mild conducted a survey of 1,429 patients with brain tumours aged between 20 and 80, and an equal number of controls with no symptoms. They found that people who used analogue phones had a 26 per cent greater chance of developing brain cancer, and that this risk rose to 77 per cent among people who had used them for more than ten years. They also found that there was a greater likelihood of the tumour occurring on the side of the head on which the phone was held, and that the most common type of tumour was the acoustic neuroma.

The researchers did not find an increased risk of brain tumours among users of digital or cordless mobile phones. However, that does not exonerate these phones. Whereas analogue phones have been used in Sweden for around 15 years, digital phones have been used only for about four. As the researchers themselves commented, a longer latency period will be needed to see whether an increased risk of brain tumours also applies to users of digital mobile-phone use. (Presented at Conference, 'Mobile Telephones and Health—the Latest Developments', London, 6–7 June 2001.)

How brain tumours might develop from EMR exposure has not yet been fully explained, though tests on phantom heads have shown that over half the radiation from a mobile phone is absorbed by the user's head. Not surprisingly, there are many reports from mobile-phone users of symptoms such as a hot ear, bleeding ears, headaches, and head pressure, confusion, memory loss, and problems concentrating.

The majority of these effects have been described in connection with use of digital mobile phones, most likely because these emit pulsed radiation which concentrates energy in short, sharp bursts, rather than the comparatively smoother waves of the older analogue phones. Because digital phones have been on the market for just a few years, and because tumours often take many years to develop, it is likely that the problems that have already emerged are only the tip of the iceberg.

New Zealand physicist Dr. Neil Cherry believes that a connection between brain tumours and mobile-phone use is likely, but will probably take decades to be confirmed. 'But the latency of cancer is decades. And so we need a large population for about two to three decades using these cell phones for a large increase in brain tumours to be observed.' (Art Bell Show, February 2000.)

Meanwhile, the writing is on the wall, with many studies already showing an association between EMR and brain tumours, and people claiming their brain tumours were caused by their mobile phones.

For more studies which have found a connection between EMR and brain tumours, see pages 185–8.

Leukemia

Together with brain tumours, the risk that has been uppermost in public awareness has been that of leukemia. In 1979 Nancy Wertheimer and Ed Leeper investigated the homes of children in Colorado who had developed leukemia. They found that twice as many deaths occurred in homes where the wiring configurations suggested a high current flow, including those close to substations and to the transformers that stepped down the voltage from 7,600 volts to 240 volts for domestic use. Risk was greater for children who had spent their entire lives at the one address. This unexpected finding was met with a hostile reception from the scientific community which doubted that there was any way that the supposedly benign electricity system could endanger people's health. (Wertheimer et al. 1979.)

In 1988 David Savitz replicated and expanded the 1979 Wertheimer study, this time in Denver. He, too, found that there was an increase in deaths from cancer/leukemia, lymphoma, and brain tumours among children living near high-current wiring from power lines. At levels above 2 mG the numbers of leukemias virtually doubled. By now the EMR-leukemia connection was looking like a genuine entity. (Savitz et al. 1988).

Since then a number of other studies have found an association between leukemia and people living in magnetic fields of just a few milligauss. In 1993 Maria Feychting and Anders Ahlbom conducted a study of half a million people who had lived within 300 metres of 220kV and 400 kV power lines. They found that children living within 50 metres of power lines had a risk of contracting leukemia 2.9 times that of the normal population. Moreover, they found that children exposed to 1 mG and 2 mG had 2.7 times the average risk of developing leukemia and those exposed to 3 mG had nearly four times the risk. There was also an increased risk for adults and the risk increased for those who had lived near powerlines for more than ten years. (Feychting et al. 1993.)

In 1997 a study was published which appeared, according to media reports, to contradict the EMR-leukemia link. This was the study by Dr. Martha Linet which once again looked at the leukemia risk for children living near powerlines. The conclusion which was widely reported was that the study had found no risk of leukemia at exposures up to 2 mG. What the press generally failed to report was that the study had found 'a tendency for the risk to be higher among subjects with summary exposure levels of 0.3 microTesla [3 mG] or more.' In other words, the study had found, not surprisingly, that minimal risk existed at 'safe' levels of exposure under 2 mG, and that an increased risk could be seen at exposures of 3 mG and above. This confirmed, rather than disproved, the work of other researchers who had found risks from powerline exposures above 2 mG. (Linet et al. 1997.)

One of the most recent—and convincing studies—to find an EMR-leukemia effect from powerline exposures is the 1999 study by Dr. Lois Green in Canada. Green found that children exposed to the highest levels of EMR had a risk of leukemia 4.5 times higher than average. Children under six years of age exposed to more than 1.4 mG had a risk 5.7 times that of children who were exposed to less than 0.3 mG. The study is significant in that it was the first to consider EMR exposure over time by providing subjects with monitors that measured radiation. (Green et al. 1999.)

Three fairly recent meta-analyses—studies which examined the data of previous research—have added weight to the association between EMR exposure and leukemia. From an analysis of fifteen studies of childhood leukemia, Dr. Daniel Wartenberg concluded that 'the overwhelming majority of studies show an increase in risk' and that the risk was quite significant, at nine times the usual leukemia rate. (Wartenberg 1998.) Similarly, an analysis of twelve studies by Dr. Sander Greenland found that leukemia risk increased at exposures of over 2 mG, and that higher exposures produced higher risk levels. Children exposed to over 6 mG had an 80 per cent greater risk of developing the disease. (Greenland et al. 2000.)

The third study, by Anders Ahlbom, analysed the data of nine previous studies comprising 3,247 children and over 10,000 controls. It found that children exposed to 4 mG or more had *double* the usual rate of leukemia. (Ahlbom et al. 2000.) According to Ahlbom, 'if we had supporting experimental data, the epidemiology would have been strong enough for a causal interpretation quite some time ago.' (*MW News* September/October 2000.)

The release of Britain's National Radiation Protection Board (NRPB) report by Sir Richard Doll in March 2001 brought the link between powerlines and leukemia into sharp public focus. As you will recall from chapter three, Doll chaired an Advisory Group for the NRPB, which reviewed a number of previous studies on powerlines and leukemia. The group concluded, 'Taken in conjunction they suggest that relatively heavy average exposures of 0.4 µT [4 mG] or more are associated with a doubling of the risk of leukemia in children under 15 years of age.' While they admitted that the evidence was 'not conclusive', they nevertheless considered that 'the possibility remains that high and prolonged time-weighted average exposure to power frequency magnetic fields can increase the risk of leukemia in children.' (NRPB 2001.)

During the period of frenzied media attention on the Doll report, and therefore substantially overlooked, another study was released which also found a connection between magnetic fields and leukemia. Conducted by German researcher, Dr. Joachim Schüz, it found that children exposed to a field of 2 mG or more averaged over 24 hours had a slightly increased risk of leukemia. However, when Schüz looked at children exposed to 2 mG or more during the night, he found they had three times the risk of leukemia. When he examined the data for children under four years of age and exposed to these fields overnight, he found they had four-and-a-half times the risk of leukemia. (Schüz et al. 2001.)

Just months later, an expert scientific working group of the International Agency for Research on Cancer (IARC) released a report which concluded that EMR from power sources is possibly carcinogenic to humans. In a press statement released on 27 June 2001, the committee wrote, 'pooled analyses of data from a number of well-conducted studies show a fairly consistent statistical association between a doubling of risk of childhood leukemia and power-frequency (50 or 60 Hz) residential ELF magnetic field strengths above 0.4 microTesla [4 milliGauss]'. This was the first time that an international agency had claimed such a connection.

Another avenue for investigating the link between EMR and leukemia has been to look at the association between leukemias and jobs with high exposure. In 1985 Dr. Samuel Milham was one of the first researchers to take this approach by investigating the death certificates of people who had been engaged in occupations assessed as having high electromagnetic exposures. He found that workers in electrical occupations had an elevated risk of leukemia and other lymphomas. (Milham 1985.)

This link has since been supported by other studies. In 1993 Birgitta Floderus of Sweden found that workers exposed to 50 Hz electric and magnetic fields had an increased rate of chronic lymphocytic leukemia and brain tumours. (Floderus et al. 1993.) The following year Dr. Giles Theriault found that the most exposed workers at electrical utilities in France and Canada had a risk of acute myeloid leukemia three times the average. (Theriault 1994.) Also in 1993 Dr. G. Matanoski found a leukemia rate up to nearly seven times the norm among telephone company workers in New York. (Matanoski et al. 1993.) In 1996 Dr. Anthony Miller found a leukemia rate for workers of a Canadian electric authority exposed to high electric and magnetic fields of 11 times the average. (Miller et al. 1996.) Interestingly, Miller found that it was the electric field—rather than the magnetic field—which correlated most with leukemia risk.

In 1997 Dr. Maria Feychting of Sweden investigated the exposure of 548 subjects with leukemias or a tumour of the central nervous system. She found that people in residences with exposures of over 2 mG had 30 per cent more chance of developing leukemia than those with low exposures at home. People exposed to the highest fields at work had 70 per cent more chance of developing leukemia than those working in low fields. Finally, people with high exposures at both home *and* at work had 3.7 times the average risk of developing leukemia. (Feychting et al. 1997.)

A number of studies have found that people exposed to EMR from communications sources also have a higher risk of developing leukemia. In 1988 Dr. Samuel Milham found that amateur radio operators had a higher rate of death from leukemia. (Milham 1998.) In 1996 Dr. Bruce Hocking found a higher rate of childhood leukemia within a four-kilometre radius of North Sydney's three television broadcast transmitters. (Hocking et al. 1996.) While several other studies have also found higher rates of leukemia among people living near broadcast transmitters, some have not. In 2000 Dr. Hocking published a follow-up study which found that children with leukemia who lived nearest to the TV towers had decreased survival rates, compared to those who lived further away. (Hocking et al. 2000.)

These are but a few of the studies that have found an EMR link to leukemia. Other can be found on pages 188–93.

Breast cancer

Leukemia is just one of the many cancers to be linked with EMR. As breast cancer reaches epidemic proportions—now affecting one in nine women—there is mounting evidence that it, too, is associated with EMR. Many studies have found that women—and sometimes men—with high exposure have an increased risk of developing breast cancers.

Several studies have found that people exposed to high fields from power sources have a higher risk of breast cancer. P. Vena found an increased rate of breast cancer among women who regularly used electric blankets at night. (Vena et al. 1991.) P. Coogan found an increased risk of breast cancer among women exposed at work (Coogan et al. 1996), as did several other studies, particularly those which investigated the risk of men and women working in electrical occupations. In a small study, M. Feychting found that women under 50 years of age exposed to 2 mG or more at work had an 80 per cent greater risk of breast cancer than those exposed to under 1 mG. (Feychting et al. 1998.)

People working in occupations where they were exposed to EMR from communications sources also have been found to have a higher rate of breast cancer. T. Tynes found an increased incidence of the disease in male radio operators (Tynes et al. 1996), and P. Demers found that men with breast cancer were more likely to have worked in electrical occupations than not. (Demers et al. 1991.) Several other researchers have also found an increased risk for the disease as a result of occupational exposure.

Some studies have examined the effects of EMR on breast cancer cells. Dr. Carl Blackman exposed human breast cancer cells to a range of conditions. He found that cells treated with the hormone melatonin began to diminish within seven days. However, in cells treated with melatonin and exposed to a 12 mG field of 60 Hz there was no such reduction. (Blackman 1989.) In other words, EMR was negating the successful efforts of melatonin to destroy the breast cancer cells. Studies such as these help to explain the association of EMR with breast cancer and how it may be contributing to this increasingly common disease.

The finding of increased breast cancers among exposed workers and a mechanism to explain the effects is a fairly compelling combination. In 1997 ten speakers at a US conference on 'EMF and light-at-night' issued a press release which stated, 'According to an international panel, electromagnetic fields and environmental light may be considered potential risk factors for breast cancer, based on existing scientific evidence.' (*MW News* November./December 1997.)

For more studies which have found a connection between EMR and breast cancer, see pages 194–5.

Other cancers

It is not only cancer of the breast that has been associated with EMR. In 1986 Lennart Tomenius found that children with cancer were more likely to have lived in homes with magnetic fields above 3 mG. (Tomenius 1986.) Birgitta Floderus, in 1994, found that railway workers exposed to high electromagnetic fields had an increased rate of cancer. (Floderus et al. 1994.) David Savitz conducted a study of children who had been diagnosed with any form of cancer between 1976 and 1983. He found that children exposed to above 2 mG had a higher incidence of cancers, leukemias, lymphomas, and soft-tissue sarcomas, and that there seemed to be a relationship between the wire codes that had been used by Wertheimer and Leeper and the incidence of cancer. (Savitz et al. 1988.)

Wertheimer and Leeper, themselves, found further associations between cancer and high exposures from the power system. In 1982 they found that adults exposed to wiring patterns typical of high-current flows had a greater risk of cancer. (Wertheimer et al. 1982.) Later they found an increased risk of cancer among children living in homes where high fields were being conducted through the water pipes. (Wertheimer et al. 1995.)

Over fifty studies have now shown that there is a statistically significant increased risk of cancer among people who work in occupations that are thought to have higher than average exposure to EMR. (NIEHS and US Dept. of Energy 1995.)

One of these is highway-patrol policemen. In 1998 Dr. Murray Finkelstein found that police officers in Ontario had a 30 per cent higher incidence of testicular cancer and a 45 per cent higher incidence of melanoma than the general population, even though the police department's overall cancer rates were lower than average. While Finkelstein thought that the melanomas might be a coincidence, he was concerned that the testicular cancer might be related to use of radar guns. (Finkelstein 1998.) So have some of the users of radar guns, themselves. When in 1996 US policeman Franklin Chappell convinced the Virginia Workers' Compensation Commission that his testicular cancer was caused by his hand-held radar gun, the city compensated him by contributing to his medical expenses. (*MW News* March/April 1996.) According to a list compiled by US highway patrolman Gary Poynter, over 200 policemen who used radar guns later developed testicular cancers or melanomas. (*MW News* November./December 1997.)

Not surprisingly, electrical workers have been found to have a high rate of cancers, with many studies finding an increased risk of brain cancer and leukemia in particular. In 1999 Dr. Bu-Tian Ji of China also found an association with pancreatic cancer. Ji found that electricians had over seven times the normal risk of developing the disease, and that the risk increased to nine times the average among men working in the industry for over 35 years. (Ji et al. 1999.)

Looking at the types of cancers developed by highly exposed workers suggested to one Swedish researcher a new explanation for the association. In late 1999, Dr. Birgitta Floderus published the results of a study of nearly two-and-a-half million men and women with cancer whom she classified according to their exposure to EMR at work. She found people in medium- and high-exposure groups had roughly a 10 per cent increased risk of cancer. Among the men there was an increased risk of cancer of the colon, biliary passages, liver, larynx, lung, testis, kidney, urinary organs, melanomas, and brain cancers, with the most obvious association being testicular cancer for young men. Women had an increased risk of cancer of the lung, breast, uterus, and ovaries as well as melanomas and chronic lymphocytic leukemia, with the most obvious connection being cancer of the uterus.

Was there a connection between these apparently diverse findings? The two cancers that correlated most highly with exposure—cancers of the uterus and testes—involve hormones of the reproductive system. Similarly, cancers of the breast and liver, and melanomas, are hormone-related. This suggests that, at least for some of the cancers, EMR may be interacting with the endocrine system. (Floderus 1999.)

Taken together, all these studies provide a strong case for arguing that EMR is a carcinogen. New Zealand physicist Dr. Neil Cherry believes that EMR is most certainly carcinogenic, and that if it were a chemical it would long ago have been labeled as such. 'There are standard techniques for assessing the carcinogenicity of chemical substances, involving cell line studies, laboratory animals, and human epidemiology. If EMR was treated in the same way it would have been declared a human carcinogen many years ago. EMR neoplastically transforms cells, causes cancer in mice, is found to increase cancer in exposed electrical workers and military personnel and in residential populations.' Whereas, as Cherry points out, the chemical carcinogen benzene may only be used at levels that are one thousand times lower than the level at which it has been found to cause effects, EMR is permitted at levels that are thousands of times greater than those where effects have been observed. (Cherry 1999.)

Other researchers believe that EMR does not cause cancer directly, but that it may promote the disease in cells that are already precancerous. There is evidence to support this point of view. Several researchers have given a batch of mice a dose of a carcinogenic chemical and then exposed them to EMR. This combination proved to be a particularly lethal cocktail, as those who received this treatment tended to develop more and larger tumours than those who received only the carcinogen *or* the EMR exposure (see chapter six).

Official attempts to classify EMR as a carcinogen have met with strong resistance. In 1998 a committee of the US National Institute of Environmental Health Sciences (NIEHS) voted that extremely low frequency electromagnetic fields should be considered to be 'possible human carcinogens', Group 2B. (Other agents classified in this group include carbon tetrachloride, chloroform, DDT, lead, PBBs, saccharin.) (*MW News* July/August 1998.) This recommendation was, however, ignored in the final report of the NIEHS.

Some years earlier, in 1994, the USA's Environment Protection Agency produced a draft report on EMR which concluded that fields from power sources were a risk factor for childhood cancer. 'The childhood cancer epidemiology studies,' it stated, 'consistently show repeated findings of a small excess relative risk of leukemia and brain cancer in children who live in homes near the electrical power distribution network.' Further, 'A large number of occupational studies have shown excess risk of leukemia and central nervous system cancers in people occupationally exposed to electromagnetic fields.' Did the report lead to legislative changes, to the allocation of more research funding, and to increased public awareness? No. It was suppressed by the EPA, and only became public some years later when it was leaked to the press by an EPA staff member. (*MW News* January/February 1998.)

Further information about the link between EMR and cancer can be found on pages 195–7.

Heart problems

Another serious condition that has been linked to EMR is heart disease. Research has shown that exposure to EMR is linked to heart-rate variability, blood pressure problems, acute myocardial infarction, and abnormal patterns that can be detected with an electrocardiograph (ECG). The ability of the ECG to register signals from the heart shows that this organ discharges fields; in fact, these fields have been detected—using a SQUID (super-conducting quantum interference) device—at the amazing distance of five metres from the body. As a super-sensitive transmitter, it is not surprising that the heart could be subject to interference from external fields.

Several studies have found an association between heart problems and high fields from the power system. A. Bortkiewicz found that electrical workers showed patterns of disturbance in ECG measurements. (Bortkiewicz et al. 1998.) D. Savitz found that people in jobs with high magnetic fields had an increased risk of death from various heart problems. (Savitz et al. 1999.) C. Ventura found that cells exposed to pulsed magnetic fields developed changes to opioid genes, which may affect the cardiac system. (Ventura et al. 2000.)

EMR from communications sources has also been associated with heart problems. Bortkiewicz found that workers in broadcast stations had problems regulating their cardiovascular function; in another study, he found abnormalities in ECG patterns. (Bortkiewicz et al. 1996 and 1997.) When S. Lu (1999) exposed rats for six minutes to ultrawide-band pulses typical of radar, he found a significant decrease in their arterial blood pressure. S. Szmigielski also found blood pressure changes and changes to heart rate. (Szmigielski et al. 1998.)

As we saw in chapter four, EMR has also been shown to have a more indirect effect on the heart, by interfering with the operation of cardiac pacemakers.

For more studies which have found a connection between EMR and heart problems, see pages 197–9.

Reproduction

The effects of radar on the reproductive system have been suspected for many years. During World War II, for example, sailors were known to willingly expose themselves to their ships' radar in order to enjoy its contraceptive advantages during shore leave. Now there is scientific evidence to support a whole range of reproductive problems—not just from radar, but from other radiofrequency and powerline exposures.

Nancy Wertheimer and Ed Leeper, the US collaborators who first identified the link between the power system and leukemia, conducted two studies to investigate whether the power system might also be responsible for miscarriages. In 1986 they found that families who used electric blankets or heated waterbeds had a higher risk of miscarriage in colder months when the heating was most likely to have been turned on. They did not find a similar pattern in families which did not use heated beds, and this suggested that the miscarriages were associated with high levels of EMR. (Wertheimer et al. 1986.) Three years later they published results of a similar study. This time they found that families had a higher rate of miscarriage in colder months, at a time when they were more likely to be exposed to EMR from a variety of sources of heating, than in the warmer months. (Wertheimer et al. 1989.)

Other researchers have found similar problems. J. Juutilainen found that women who had miscarried were more likely to have lived in a home where there were high fields (Juutilainen et al. 1993), while K. Belanger found that women who used electric blankets at the time of conception had an increased risk of miscarriage. (Belanger et al. 1998.) In another study, D. Li found that women who had experienced difficulty conceiving and had used electric blankets during pregnancy had a more than four-fold risk of producing children with problems of the urinary tract. (Li 2001.) Other studies have also found that parents exposed to high fields produced children with deformities or health problems.

Whereas most research has examined the reproductive risks of magnetic fields averaged over a period of time, an innovative study by Dr. De-Kun Li (2001) looked at the risks posed by peak magnetic fields. Dr. Li monitored 969 women in the early stages of pregnancy and their typical exposure. He found that women exposed to a maximum magnetic field of 16 mG (a figure that seemed to be a threshold for effects) and over had 80 per cent more miscarriages than the less-exposed volunteers. The possibility that magnetic peaks may be affecting health has serious implications for future research.

Several studies have shown that using a VDU during pregnancy increases the risk of miscarriages and birth defects, although others have failed to confirm these findings.

M. Goldhaber tracked the pregnancies of 1,583 US women in the early 1980s. She found that women who used VDUs for more than 20 hours a week in the first three months of pregnancy had a higher risk of miscarriage than women working in occupations where they did not use VDUs. (Goldhaber et al. 1988.) Studies by M. Lindbohm and M. McDiarmid in the early 1990s also found increased miscarriage rates among VDU users. Perhaps one reason why some studies have failed to find a link between computer use and miscarriages is the variability in emissions of different types of computers. In recent years most computer manufacturers have voluntarily adopted the strict Swiss limits which expose operators to lower emissions than previously.

Exposure to microwave radiation has also been linked with reproductive problems. Dr. John Goldsmith investigated the health of people who had been exposed to microwaves in military occupations, people who lived near a radar station in Latvia or a television station in Hawaii, female physiotherapists who used short-wave

No chance of me getting a brain tumour, mate.

and microwave diathermy, and embassy staff exposed to microwaves in Russia. He found that these people had an increased risk of a number of health problems, including reproductive problems such as miscarriages. (Goldsmith 1995.)

Not all reproductive problems seem to be caused by exposure of the foetus. There is some evidence that microwaves affect the sperm, itself. T. Weyandt found that artillerymen in the US army whose work was likely to expose them to microwaves, had lower sperm counts, and ejaculated lower levels of sperm than unexposed men. (Weyandt et al. 1996.) Several studies have also found testicular changes in mice exposed to EMR from microwaves.

Cot death (SIDS)

Given that EMR has been associated with risk to foetuses, it will come as no surprise that it has also been linked with cot death. Pioneer EMR researcher Robert Becker describes, in his book *Cross Currents*, a possible melatonin connection to SIDS. As you will read in the following chapter, exposure to EMR reduces the levels of the hormone melatonin, which is secreted by the pineal gland and helps to regulate the body's natural rhythms and attacks cancer-producing free radicals. Rhode Island medical examiner Dr. William Sturner found that SIDS victims had much lower levels of melatonin than babies who had died of other causes, and this suggested they had received high EMR exposures. According to Becker, the low melatonin levels may have depressed the infants' respiratory controls so much that they stopped breathing altogether.

Becker also reports a connection with radiofrequency exposures. Dr. Cornelia O'Leary, in her investigations of SIDS, found that eight deaths occurred on a single weekend (four of them within a single period of two hours) when a radar unit was being tested just seven miles away. (Becker 1991:259–60.) While reports such as these are hardly conclusive, they certainly suggest a possibly fruitful avenue for future SIDS research and the merits, meanwhile, of reducing exposures at home.

For more studies which have found a connection between EMR and reproductive problems, see pages 199–203.

Brain function

The brain is a highly complex dynamo, a human computer in touch with every cell of its body, constantly receiving and transmitting messages. It contains around 100 billion cells (neurons) each with numerous tentacles reaching out to other cells, forming a complex network of around 100 trillion connections. Each neuron fires an electrical impulse about once every five seconds, and these signals can be measured and recorded by an EEG (electroencephalogram) machine which is routinely used by the medical profession to diagnose neurological conditions.

The brain is influenced by extremely small levels of EMR. In an eye that has been adapted to the dark, for example, a cell can be stimulated to fire by just one photon of light. (A cell phone puts out billions of photons per second.) Similarly, just turning on a domestic light can change brain-wave patterns.

Not surprisingly, many studies have found that EMR affects the electrical activity of the brain. Even static fields and very low frequencies (for example, 1.5 Hz, 10 Hz, and 75 Hz) have produced changes in brain patterns on EEG recordings. 50 Hz fields from the power system and radiofrequency signals also produce effects, and these can last longer, sometimes several hours longer, than the exposure itself. L. von Klitzing (1995) exposed volunteers to EMR from mobile phones, and noticed that changes were most apparent in the alpha range of brain waves. G. Freude found that mobile-phone exposures caused a decrease in waves known as slow brain potentials. (Freude et al. 2000.) Some studies have showed that exposures before sleep affect the quality of sleep and that, in some cases, sleep can actually be improved.

The significant question to emerge from these studies is whether or just how much these changes affect people's health. While the answer is not yet clear cut, there is certainly evidence that EMR is affecting a number of systems associated with the brain.

A more detailed list of studies which have shown that EMR affects the electrical activity of the brain can be found on pages 203–5.

The nervous system

In 1969 Jose Delgado found that, by implanting electrodes in particular sites of the brain, he was able to use electricity to affect people's moods. Delgado found that, by directly stimulating sites of the brain in this way, he could produce fear, anxiety, euphoria, rage, pleasure, or even major personality changes; this has been confirmed by other researchers since. There is evidence that EMR is doing much the same thing indirectly. Over the last twenty years, a good deal of evidence has accumulated that EMR can also produce a range of effects on the nervous system, from tiredness to depression and suicide.

One of the symptoms most often associated with exposure to high fields from the power system is depression. Three significant studies were undertaken by British GP Stephen Perry, who was concerned about the inexplicably high level of depressive symptoms among patients he encountered. In 1981 Perry found that people who lived in high magnetic fields next to powerlines were 40 per cent more likely to commit suicide. (Perry et al. 1981.) Next he found that, in blocks of flats with underfloor or electric storage heating systems, people who lived near the main cable had a higher rate of depression. (Perry and Pearl 1988.) Finally, in 1989, he found that patients admitted to hospitals for depressive illnesses were more likely to have lived in higher residential magnetic fields. (Perry et al. 1989.)

Other studies have also found that people living in high fields from power sources have a higher risk of depression. Working with Stephen Perry, Maria Reichmanis established a correlation between the location of electricity transmission lines and higher-than-expected suicide rates in parts of England's Midlands. (Reichmanis et al. 1979.) C. Poole found that people who lived near powerlines reported a higher number of symptoms associated with depression (Poole et al. 1993), and P. Verkasalo found that there was 4.7 times the risk of severe depression if people lived within 100 metres of a high-voltage line. (Verkasalo et al. 1997.) L. Bonhomme-Faivre found that people who had worked in high fields for between one and five years showed an increased rate of several symptoms, including depression, irritability, and melancholy. (Bonhomme-Faivre et al. 1998.) Van Wijngaarden found that electricians had double the normal rate of suicide and that electrical linesmen had 1.5 times the normal rate, and that the risk of suicide increased with higher exposure to power fields. (van Wijngaarden et al. 2000.)

Radiofrequency radiation also seems to affect the nervous system. People in the Swiss village of Schwarzenburg, not far from a radar station, reported a range of symptoms that their neighbours further from the radar station were not experiencing. These included sleep problems, headaches, fatigue, irritability, pain in the lower back, and limb pain. Children from the village also performed worse at school than those living over four kilometres away. (Alpeter et al. 1995.)

But what of the effects of mobile-phone use on the nervous system? A number of studies have reported a range of symptoms experienced by mobile-phone users. Dr. Bruce Hocking reported that mobile-phone users were experiencing symptoms of burning or a dull ache in their heads. (Hocking 1998.) Other researchers have also identified symptoms of headaches, feelings of warmth, or burning and fatigue.

Dr. Hocking recently reported an extremely interesting case study of a gentleman who had been a long-term mobile-phone user. When he used his mobile phone in his right hand, the man noticed 'unusual brief sensations on the right scalp'. Then, after two conversations on his mobile phone, each lasting for nearly an hour, he 'developed persistent symptoms' on his scalp, cheek and neck. (Hocking et al. 2000.) Describing this case to the Senate inquiry on EMR, Dr. Hocking said, 'This is the first time that I am aware of that there has been a clear demonstration of a health effect in humans attributable to a mobile phone. I agree it is only one case … Nonetheless, I think it is a significant warning when you see it in context with the previous 40 cases that I was reporting that were getting similar sorts of symptoms that there is considerable likelihood that mobile phones, at the low levels of radiofrequency which they operate on, are causing disturbances of neural function.' (*Hansard* 22 September 2000.)

There are many additional studies which have found that EMR affects the nervous system, some of which are described on pages 205-9.

'Not tonight, honey. I've got a headache.'

Neurodegenerative diseases

Several neurodegenerative diseases have been associated with exposure to EMR. Most studies have examined occupations with high exposure, and have found an excess of diseases such as Alzheimer's and Lou Gehrig's disease (Amyotrophic

Lateral Sclerosis), and there are suggestions that EMR may also affect Parkinson's disease. As is the case for cancer, these diseases are multi-causal, and damage accumulates over a lifetime.

Eugene Sobel investigated the incidence of Alzheimer's Disease among people who were exposed to high levels of EMR from the power system at work. He found that people working as dressmakers, seamstresses, and tailors, who were exposed to higher than average fields, had three times the risk of the disease. In a later study, he found that people with Alzheimer's were more likely to have worked in jobs with high exposure. A number of other studies have also shown that people with high work exposure had a greater risk of Alzheimer's Disease, ALS, and dementia.

EMR may be contributing to brain diseases of these sorts by diminishing the brain's two vital defence systems. The first of these is the blood-brain barrier. EMR's ability to cause breaches of this barrier is one of its best-researched and most widely accepted effects. Because the barrier exists to protect the brain from incursions by harmful chemicals and certain immune-response cells, breaches are extremely serious and can lead to neurodegenerative diseases.

In late-1999 Swedish researcher Leif Salford described to the press how breaches in the blood-brain barrier could be caused by exposure to low levels of microwave radiation from a mobile phone. In fact, the breaches occurred at the extremely low levels of radiation that would be encountered near a mobile-phone tower or simply standing near someone using a mobile phone. (*Svenska Dagbladet* 16 September 1999.) This was not a chance finding. There are numerous other studies that have found that EMR causes breaches to the blood-brain barrier.

The brain is not the only organ to have a protective barrier such as this. The testicles have a barrier and so does the thymus (which generates T-cells) to protect stem cells from being devoured by the body's immune system. Currently research is taking place to see whether EMR is also breaching these barriers.

Just in case harmful free-radicals do successfully negotiate the blood-brain barrier, a second line of defence exists. This is provided by the hormone melatonin, a free-radical scavenger which can pass through the blood-brain barrier without impediment. Unfortunately, however, melatonin is also affected by exposure to EMR. Quite a number of studies have shown that people who are exposed to high fields, either during the day or at night, have reduced melatonin levels (see chapter six). This may help to explain not just the link with neurodegenerative diseases, but with a host of other problems as well.

Additional details about studies that have found a connection between EMR and neurodegenerative diseases are on pages 209–210.

Learning and performance

Hundreds of rats in dozens of laboratories have scampered through mazes and floundered through tanks of water to help answer researchers' questions about whether EMR affects learning and performance. For many rats it has.

Dr. Henry Lai tested the memory and performance of rats in a series of water trials. Before each training session some rats were exposed to a 60 Hz magnetic field, others were given a mock exposure, while a third group was not exposed at all. During the trials, the rats were deposited into a tank of water containing a submerged platform which they soon learned to locate. Lai noticed that the rats that had been exposed to the magnetic field were just as successful as the other groups in learning to locate the platform, but their swimming speed was much slower. In the final trial, Lai deposited each group into the water tank, this time with the platform removed, and noted whether the rats were looking for the platform in the appropriate location or whether they appeared to be 'lost'. This time the rats that had been exposed to the magnetic fields spent less time in the area where the platform had been, and showed quite different swimming patterns to their comrades in the other groups. From this he concluded that the magnetic-field exposure had affected the 'spatial' memory of the rats. (Lai et al. 1998.)

Two years later, Lai repeated this experiment with a colleague, B. Wang, exposing mice to pulsed EMR of the sort used in microwave ovens rather than the power system. This time he found that rats that had been exposed to microwaves were just as quick at swimming as their sports mates but took longer to learn to locate the submerged platform. In the trial without the platform, the exposed rats again spent more time looking for the platform in the wrong places, and showed different swimming patterns. This suggested that the microwaves had also affected the spatial memory of the rats. (Wang et al. 2000.)

Other researchers have found similar results. K. Salzinger (1987) found that rats that had been exposed to high fields from the power system as foetuses later had problems learning and made more mistakes than rats who had not been exposed in this way. Z. Sienkiewicz found that rats exposed to a 50 Hz magnetic field were slower at learning than unexposed rats. (Sienkiewicz et al. 1998.) J. Wu found that rats exposed to radiofrequency signals had more problems learning a task, and attributed this to changes in the neurotransmitters in parts of their brains. (Wu et al. 1999.)

Valuable as these experiments may be, they do not resolve the question of whether EMR affects the learning and performance of humans. Studies that have investigated this question have yielded some ambivalent results. M. Trimmel found that people exposed to 50 Hz fields had reduced performance in the areas of attention, perception, and memory. (Trimmel et al. 1998.) A. Kolodynski found that children who lived near the Latvian radar station of Skrunda had reduced memory, attention, reaction times, motor function, and neuromuscular function. (Kolodynski et al. 1996.) L. Duan found that workers exposed to high-frequency radiation performed worse in tests to assess their neurobehavioural function than controls. (Duan et al. 1998.)

On the other hand, some studies have found that exposure to radiation from a mobile phone actually improves performance. Dr. A. Preece found that students scored better in the speed of their yes/no responses to a task after exposure. (Preece et al. 1999.) Similarly, M. Koivisto found that volunteers responded more quickly to tests gauging reaction times and vigilance, and needed less time to perform mental arithmetic. (Koivisto et al. 2000.)

These results should not be read as a justification for increasing mobile-phone use. They show that the brain has indeed been affected by radiation from the phone, and it could well be many years before we understand the long-term consequences of this effect.

A more complete list of studies connecting EMR with learning and performance in on pages 211–13.

Sleep

There is evidence that exposure to EMR from power and radiofrequency sources affects the quality and quantity of sleep. T. Akerstedt compared the sleep of 18 volunteers before and after exposure to a magnetic field of 10 mG. He found that this exposure reduced the total amount of sleep time, sleep efficiency, and slow-wave activity of the brain. (Akerstedt et al. 1999.)

Dr. R. Huber exposed male volunteers to the radiation from a mobile phone for 30 minutes just before sleep. This had the effect of changing the electrical signals of the brain in some frequencies during later sleep. In other words, use of a mobile phone affects brain waves and sleep even beyond the period of exposure. (Huber et al. 2000.)

The effect of EMR on sleep may be far more than inconvenient; it may affect memory and learning. Dr. C Graham showed that high magnetic fields from the power system adversely affected the sleep of volunteers. They spent more time in Stage II sleep and less in REM (Rapid Eye Movement) sleep, and also reported sleeping less well and feeling less rested. (Graham et al. 1999). A study by K. Mann and J. Roschke exposed volunteers to the radiation from a mobile phone. It found that the subjects went to sleep more quickly, had changed brain-wave patterns, and had less REM sleep. (Mann et al. 1996.) As REM sleep is associated with memory and learning, disturbance of this phase of sleep may affect people's performance. It would be interesting to know the effects of regular mobile-phone use on people's sleep patterns and performance.

For more details about studies that have found a link between EMR and sleep problems, see page 213.

The immune system

The immune system is the body's defence against foreign intruders: it consists of the lymphatic system, the tonsils, and the white blood cells. Any stressor which impairs its function could trigger a whole range of effects, depending on individual susceptibility.

One of the significant studies which found that EMR affects the immune system was led by Dr. Michael Repacholi at Adelaide Hospital. Repacholi's team exposed groups of transgenic mice—mice especially bred to be susceptible to environmental changes—to the radiation from a GSM mobile phone operating at 900 MHz for one hour per day for a nine to 18 month period. They found that the exposed mice had nearly two-and-a-half times as many lymphomas—cancers of the immune system—as unexposed control mice. (Repacholi et al. 1997.)

What makes this study particularly important is the connection it has shown between EMR and B-cell lymphomas. Repacholi found that the radiation exposure caused the mice to develop not the T-cell lymphomas that were expected in this strain of mouse, but B-cell lymphomas that are involved in around 85 per cent of cancers. B-cells play an important role in the body's immune system by producing antibodies and detecting cancer cells.

There are many additional studies which show a connection between EMR and the immune system; they are outlined on pages 214–16.

Hormones

Closely allied to the immune system is the endocrine system, which controls our hormones, and which is also affected by exposure to EMR. The endocrine system is one of the body's two major control systems: it is responsible for transmitting chemical messages—hormones—through the blood stream. Hormones can stimulate or inhibit body processes such as ovulation, the flight-or-fight response, the rate of metabolism, and body rhythms. The endocrine system includes glands such as the pineal, pituitary, thyroid, thymus, adrenals, pancreas, ovaries, and testes.

The pineal gland produces the hormone melatonin which has a vitally important role in maintaining health. Melatonin contributes to healthy sleep by lowering body temperature and helping to maintain sleep. It reduces cholesterol, thereby reducing the risk of heart disease. It also reduces blood pressure and blood clotting, thereby reducing the risk of strokes. Importantly, melatonin scavenges free radicals which contribute not only to cancers, but also to neurodegenerative diseases. Melatonin also plays an important role in the immune system, where it interacts with the T-cells to set in train a process that stimulates a number of immune cells, including NK cells which attack cancers and viruses. (Reiter and Robinson 1995.) Finally, melatonin helps to promote the growth of bones. (Roth et al. 1999.)

Many studies have shown that exposure to EMR reduces the production of melatonin. This has been suggested as a mechanism to explain the excess of cancers among exposed people (see chapter six).

The pancreas produces the hormone insulin, which is important for regulating glucose metabolism. Lack of insulin results in diabetes; so, if the pancreas is affected by EMR, we could reasonably expect to find an increase in diabetes in highly exposed people. There is some evidence for this association. Dr. Ivan Beale conducted a study of people living near high-voltage powerlines in New Zealand. He found, among other effects, that there was an increase in type II diabetes. (Beale et al. 1997.) In another study, J. Bielski found that the majority of workers exposed to radio waves showed abnormal blood-sugar levels after being given a dose of glucose. (Bielslki et al. 1996.)

If the production of insulin is affected by EMR, as seems likely, we can better understand the alarming escalation of this condition coinciding with the steady increase in power consumption and the proliferation of telecommunications systems in society. A recent study by the International Diabetes Institute found that the number of people with diabetes has doubled in the last 20 years, and that one in four adult Australians now has either diabetes or problems with sugar metabolism. (Health Communications Australia 2000.)

Further evidence that EMR affects the endocrine system comes from the study by Floderus described earlier. She found that the prevalence of testicular and ovarian cancers developed by people working in high EMR environments suggested that the endocrine system was involved. (Floderus et al. 1999.)

Summaries of studies relating to hormonal effects of EMR are on pages 216–18. See also studies on EMR suppression of melatonin, chapter six.

Allergies to EMR

Another health problem to have arisen in the last few decades is the allergic response to EMR. People with allergies of this sort experience health problems in varying degrees when they are exposed to EMR, with different frequencies eliciting different responses. For some people, exposure to power frequencies when walking near powerlines, shopping in large complexes, travelling by train, or driving a car causes anything from mild reaction to extreme illness. Other people are unable to tolerate radiofrequency signals generated by mobile phones or even phone towers. Naturally, this makes living in modern cities an almost impossible challenge.

Common symptoms experienced by allergy sufferers include irritability, lack of energy, inability to concentrate, hyperactivity, tiredness, and sleeping problems. Because some of these symptoms are neurological, the condition has sometimes been dismissed as 'in the mind'. However, more and more evidence is accumulating that allergies to EMR constitute a definite clinical condition. Dr. Olle Johansson has shown, for example, that simply watching TV causes changes to the skin of people with allergies to EMR. (Johansson 1994.)

There is a close connection between allergies to EMR and allergies to chemicals. Not only are the symptoms virtually identical, but people with chemical allergies often develop allergies to EMR—and vice versa. A study of fifty people with chemical allergies who were given a neuropsychological screening test found that most had some degree of brain-function impairment, and that the effects ranged from mild to severe. Tasks that the subjects found difficult included following a sequence of instructions involving looking, listening, remembering, and performing, verbal recall, gross motor movements, and spatial organisation. They also had concentration problems, difficulty remembering what they had seen and reproducing it, and understanding verbal statements. This study raises the question of whether people with EMR allergies experience impairment of brain function, and suggests that the symptoms of people with environmental allergies are far from psychological. (Butler et al. 1977.)

Cancer, leukemia, brain tumours, nervous system disorders, immune dysfunction, genetic damage, learning difficulties, and behaviour problems ... This is a formidable battery of health problems to be associated with an environmental pollutant.

And this is but a tiny fraction of the studies that have found health problems from EMR exposure. There are many interactions at a cellular and genetic level, as we will see in the following chapter.

-6-

How does EMR affect the body?

Hundreds of studies and the experiences of countless people can't all be wrong. There is now a compelling and rapidly growing body of evidence that EMR can have a deleterious effect on the human body. But if this is the case, how does it do so?

In order to understand how EMR affects the body, it is first necessary to recognise that the body is, in fact, a magnificently designed electromagnetic instrument. Every atom of every cell of every organ within it has electrical components. The nucleus is neutral, protons are positively charged, and electrons are negatively charged. Atoms that have lost or gained an electron in their outer shell are called ions, and have an electrical charge as they search for another atom with which to bond to fill their electrical hole, so to speak.

Every cell in our body is an electro-chemical factory. Inside the cell are positively charged potassium ions, while outside are positively charged sodium ions and negatively charged chloride ions. When sodium ions flow through the channels in the cell membrane, they cause a change in the voltage of the membrane of 10 to 100 millivolts (0.01–0.1 volts) in less than one-thousandth of a second. It is this action that stimulates the release of chemical messengers to produce a vast range of responses in the body. Even the tiny follicles of our hair emit a magnetic field, which can now be measured using SQUID technology.

The larger organs, structures, and systems of the body, it follows, also have electromagnetic characteristics. Our organs resonate at specific frequencies. Our bones generate piezoelectricity, and fractures result in an increased flow of negative current around the site of injury. Blood, which contains small amounts of iron, is semi-conductive, as are our tissues and our DNA. Our hearts emit electromagnetic waves, which can be measured using an electrocardiogram (ECG). So do our brains, with deep sleep producing delta waves at 0.5–3 Hz; light sleep producing theta waves at 4–8 Hz; meditation or relaxation producing alpha waves at 8–14 Hz, and everyday awareness producing beta waves at 14–35 Hz—all measurable using an electroencephalogram (EEG).

It seems that the body is equipped with a labyrinth of pathways that allow electromagnetic messages to circulate throughout it. According to veteran EMR researcher Dr. Ross Adey, 'Cells in tissue are separated by narrow fluid channels, or 'gutters'... that act as windows on the electrochemical world surrounding each cell.' Through these channels flow electromagnetic messages from both within the body and outside it. The channels also contain electrochemical antennae made of protein that protrude from the cell membranes. (Adey et al. 1996.)

Given its electromagnetic design, it is hardly surprising that the human body is exquisitely sensitive to external fields. Light, which is part of the electromagnetic spectrum, has a critical effect on our bodies. It affects timing, duration, and need for sleep, it regulates body rhythms, influences chemicals that stimulate weight gain and loss, and affects our mood and our behaviour. Our eyes are able to translate light into colour, and our skins to convert sunlight into Vitamin D. We also respond favourably to some sounds—which, after all, are merely vibrations—and not to others. Our pineal gland contains traces of magnetite by which we orient to the magnetic field of the earth.

One experiment actually measured how the body responded to an external field. A. Tinniswood exposed the body to a number of external stimuli to the visual, touch, and auditory senses. He found that an auditory stimulus caused the brain to produce an average electric field of 19.2 µV/m with a current density of 0.3 µAmps per square metre. (Biolectromagnetics Society (BEMS) meeting, Abstract 7-2 1999.)

One of the reasons for the body's sensitivity to external fields is the fact that it absorbs radiation. This can be experienced often when tuning a radio or television set. When you stand next to the set, the signal is clear, but as soon as you move away it deteriorates. This is because your body was conducting the signal to the set, rather like an aerial. A more dangerous but well-known example of this effect occurs when a fluorescent light tube, held by a person standing below a high-voltage powerline, lights up.

These are not isolated incidents. The body is constantly picking up signals from the environment or, to put it another way, currents are constantly being induced in the body, and this is recognised by authorities worldwide.

Currents are induced inside the body from EMR from any electrical source—such as a powerline, a hairdryer, a mobile phone or a phone tower. Any external electric field will cause an electrical charge to travel through the body. In the same way, an external magnetic field induces a magnetic field in the body and this, in turn, has an effect on the electrical charges in the body. Electrical fields tend to be more easily induced over short distances than magnetic fields and have a greater impact on the body.

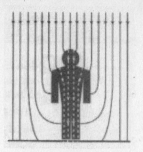

A person standing in an
electric field, showing
induced current.

A person standing in a
magnetic field, showing
induced current.

Illustration courtesy National Institute of Environmental Health Sciences, National Institutes of Health

As creatures that are electromagnetic in design and responsive to the electromagnetic signals of the natural environment, is it hardly surprising that we respond also to the electromagnetic signals that have been created artificially. Yet not all signals appear to have a harmful effect. Why?

Windows of effects

There are a number of parameters that affect how or whether the body responds to artificial EMR. These include the frequency of the signal, its power, and the amount of exposure received.

Unlike most other toxins, EMR appears to cause effects only within certain narrow bands or windows. For example, some frequencies cause problems and not others, and the same is true of power levels and amounts of exposure.

This is tremendously important. It means that science is not looking for a simple case of cause and effect, as perhaps with a disease. Nor is it looking for a conveniently linear dose-response relationship, as might be the case for a chemical (where a little arsenic, for example, is not a problem but a lot is extremely bad news). The fact that scientists need to take into consideration particular combinations of frequency, power, and length of exposure makes EMR an enormously difficult issue to study. It also helps explain why scientists have so often found apparently contradictory results, with one laboratory reporting a particular effect but another, using the same experimental design, finding none.

It also opens the possibility that EMR, at particular combinations of frequency, power level, and duration, may have beneficial effects.

Frequency

Perhaps the most obvious consideration when considering the effects of EMR is the frequency. The radiation from a powerline operating at 50 Hz is quite different in nature, for example, to the radiation from a digital mobile phone operating at a much higher frequency of 900 MHz. So it is not surprising that all frequencies do not produce the same effects.

There is certainly evidence that effects occur in certain frequency windows— that is, at some frequencies and not others. Two of the early pioneers in this field were Susan Bawin and Ross Adey, who found that the greatest outflow of calcium from animal brains occurred at a frequency of 16 Hz. (Bawin et al. 1976.)

Not all scientists were able to reproduce these results, however, and some found that the maximum effects occurred at slightly different frequencies. After trying unsuccessfully to reproduce Bawin and Adey's work, Carl Blackman came to the conclusion that calcium ion efflux was strongly influenced by the strength of the earth's magnetic field at different laboratories, a complication which explained the diverse findings.

Nevertheless, science has identified frequency windows for particular effects. For example, changes to the blood-brain barrier have been found at frequencies of 915 MHz, 1200 MHz, 1300 MHz, 2450 MHz, and 2800 MHz. Similarly, effects on calcium have been found at frequencies of 50 MHz, 147 MHz, 450 MHz, and 915 MHz. (Lai 1998.)

Power Density

Another parameter significant in eliciting an effect is the intensity—or power density—of the signal. (The power density is how much power from a source a person actually receives.) As was the case for frequency, research has found that effects occur within certain windows of power density. Carl Blackman, for example, found that the effects on calcium ions in animal brains occurred at power densities of 1.75, 3.85, 5.57, 6.82, 7.65, 7.77, and 8.82 mW/cm^2 but not at a range of other power density levels he tested. (Blackman et al. 1989.)

There is a widely held assumption that EMR operates like a toxic chemical, where higher doses produce greater effects. However, EMR does not behave like a toxic chemical. It is not a foreign substance but rather it is intrinsic to the operation of the human body. The higher-dose-the-greater-the-effect argument is quite invalid.

The opposite end of this argument, that the power is too low to be a problem, is equally invalid. Many studies have found effects at extremely low power levels.

Kwee and Raskmark (1997) found changes in the proliferation of cells at a Specific Absorption Rate (SAR) of just 0.000021–0.0021 W/Kg.* This level is minute in comparison to the level of 4 W/Kg in the standard that is legally allowed in Australia.

Quite a number of studies have found that more effects occur at lower levels of exposure than at higher levels. An Australian researcher, Dr. Peter French, exposed astrocytoma cells to EMR at 835 MHz. He found that, at the power that would be emitted by a mobile phone, there were changes to the proliferation (division) of cells; but that, at powers five times higher than this, no such changes to cell proliferation could be seen. (French et al. 1997.) Similarly, Dr. C. Daniells from the UK found that worms exposed to 750 MHz showed a greater stress response at lower powers than at higher powers. (Daniells et al. 1998.) In another study, Daniells and de Pomerai found that the stress response occurred even at the extraordinarily low specific absorption rate of 0.001 W/kg (which is very much lower than the 4 W/kg that is allowed under Australia's existing standards). (de Pomerai 2000.)

A fascinating explanation for this phenomenon has recently been advanced by Sydney's Jocelyn Laurence. Laurence suggests that microwaves operating at low power can act upon protein molecules in cells in such a way that growth, genes, and cellular communication are affected. Microwaves operating at higher power can activate a stress response by releasing 'heat shock' proteins to protect the cell. (Heat shock proteins are produced not just by heat, but also by other forms of stress.) Thus, effects at lower power slip through the body's natural defence system, and cause problems that do not occur at higher power. This theory may help explain some of the apparent anomalies that have been seen in research to date. (Laurence et al. 2000.)

It is quite likely that our bodies are particularly sensitive to low power levels because we are attuned to the low power of the natural world. It is by orienting themselves to these fields that living creatures find their way. Because the fact that more effects may occur at lower, rather than higher, power levels is the reverse of the usual more-toxins-are-worse model, it has not filtered through to industry and administrators who constantly assure us that the emissions from a particular proposal are too low to cause health problems. On the contrary, they may be just low enough to cause problems.

Allan Frey explains the effects of EMR at varying powers and frequencies beautifully, comparing the body to a radio receiver: 'An electromagnetic signal a radio detects (let us call it signal X) and transduces into the sound of music is almost unmeasurably weak. Yet the radio is immersed in a sea of EM [electromagnetic] signals from power lines and radio stations, TV stations, radars etc. The radio doesn't notice the sea of signals because they are not the appropriate frequency of

---

* *Specific Absorption Rate (SAR).* This is the rate at which microwaves are absorbed by a mass, usually the body. The SAR will be affected by the size and density of the body and the frequency of the field. It is measured in watts/milliwatts or microwatts per square metre.)

modulation. Thus they don't disturb the music we hear. If we expose the radio to an appropriately tuned EM signal or harmonic, however, even if it is very weak compared to signal X, it will interfere with the music. Similarly, if we expose a (human being) living system to a very weak EM signal, if the signal is appropriately "tuned" it could facilitate or interfere with normal function.' (Frey 1993.)

Amount of exposure

Clearly the amount of exposure a person receives is likely to play a significant part in how the body responds.

While governments and industry maintain that the radiation from mobile-phone base stations is too low to be a problem (which is based on a dubious assumption, as we have seen) they have not taken into consideration the fact that these devices are emitting radiation 24 hours a day, every day of the year. Drs. Lai and Carino have found evidence that exposures to 'lower intensity, longer dura-tion' sources are just as potent as exposures to 'higher intensity, shorter duration' ones. According to Lai, 'A field of a certain intensity, that exerts no effect after 45 min. of exposure, can elicit an effect when the exposure is prolonged to 90 min.' (Lai 1998.)

There is also evidence that the number of exposures is significant. Lai's studies show that the effects of exposure are not 'forgotten' after each exposure but that there is a cumulative effect. 'Definitely, DNA damage in cells is cumulative.' (Ibid.) According to Lai, the response of the central nervous system to EMR is likely to be a stress response. Stress responses are known to be cumulative, and involve initial adaptation and finally a break down of homeostasis.

One study which found that the length of exposure related to the effects experienced was conducted by Dr. Samuel Milham. Milham, who investigated the health of workers in an office located above three 12kV transformers (see chapter three), found that the risk of cancer increased with the duration of employment. While only one cancer was found in the 254 people who worked for less than two years, seven cases were found in the 156 people who had worked for two or more years. Milham believes that this indicates that the body has a repair mechanism that allows it to cope with stress for a period before succumbing to illness. (Correspondence.) Hence, long-term continuous exposure may be necessary for cancer to develop though, as Milham points out, childhood leukemia is obviously an exception.

Polarisation

Another parameter relevant to EMR's effects on the body is how it is polarised. Polarisation refers to the way that an electric or magnetic field is oriented, and it varies from linear polarisation to circular polarisation.* In the everyday environ-ment, fields tend to be polarised somewhere between these two extremes—in an elliptical pattern.

Several studies have shown that greater effects occur when the field is polarised in a circular fashion, rather than linearly, as you will read below.

Because so many studies have shown an association between EMR exposure and cancers, including leukemia, much research has sought to demonstrate a mech-anism for this effect. Evidence thus far points to several different mechanisms.

The melatonin effect

One of the most convincing and well-accepted mechanisms is that the EMR link to breast cancer is by way of an effect on the pineal gland. This tiny gland, situated near the forehead and at the base of the brain, produces hormones that allow our body's natural rhythms to synchronise with those of the earth, including the day/night cycle, and these rhythms are known as circadian rhythms. The pineal produces the hormone melatonin—among others—which, as a free-radical scav-enger, rids the body of potentially carcinogenic toxins and has been shown to reduce the growth rate of breast cancer cells.

Many studies have shown that exposure to EMR reduces melatonin levels. B. Wilson found that people sleeping in high fields from electric blankets had lower levels of melatonin, and a number of other researchers found that high expo-sures from the power system, whether at work or home, also reduced the levels of the hormone. (Wilson et al. 1990.) James Burch found that people who reported frequent use of their mobile phones had lower than normal levels of melatonin (Burch et al. 1997), and S. Afzal found that the magnetic fields from a VDU also reduced melatonin. It is likely that the reduced levels of melatonin impair the body's ability to destroy cancer cells, leading to an increased risk of breast—and other—cancers. (Afzal et al. 1996.)

Studies by James Burch and Masamichi Kato suggest that one of the impor-tant parameters for reducing melatonin is how a field is polarised. Burch investi-gated the levels of melatonin in electrical utility workers in different work environments. He found that those who worked in a substation for more than two hours per day or were exposed to three-phase powerlines—both of which had cir-cularly polarised fields—had quite low levels of melatonin at night. However,

---

*Imagine a narrow jet of water from a hose. Wave the hose up and down and you send a vertical line of waves cascading from it. Or wave the hose rapidly from side to side and you send out a horizontal line of waves. This is a little like linear polarisation. Now turn the hose to a gentle spray and watch the constant stream of circles of water flow from it. This is a little like circular polarisation.

those who worked on single-phase powerlines—which have linearly polarised fields—did not show such an obvious reduction in melatonin. From his observations, Burch hypothesised that circular or elliptical polarisation may be associated with the suppression of melatonin. (Burch et al. 2000.)

This hypothesis is consistent with the findings of the earlier study in Japan by Kato. Kato exposed rats to magnetic fields that were either linearly polarised or circularly polarised. He found that rats exposed to a field of 14 mG with circular polarisation for six weeks had reduced levels of melatonin in both their pineal glands and blood plasma. However, rats exposed to a 10 mG field with linear polarisation showed no such effects. Even at the higher exposure of 50 mG with linear polarisation, the rats showed no reduction in melatonin in their pineal glands, though levels in blood plasma were reduced. (Kato et al. 1997.)

If a field with circular polarisation has more effect on the body than a field with linear polarisation, then it is no wonder that much of our laboratory research to date has failed to find effects. In many of the multi-million dollar research programs conducted in the US by the National Institute of Environmental Health Sciences (including the RAPID research program) linearly polarised fields were used because they were both cheaper and easier to use.

Not only is melatonin a free-radical scavenger, but it plays an important role in sleep, it reduces cholesterol and blood pressure, and helps maintain the immune system. This may help explain why EMR has been implicated with heart problems, low immunity, blood pressure problems (Braune 1998), and learning problems that are associated with poor sleep.

It also seems to be the case that melatonin plays an important role in bone growth. Dr. Jerome Roth and colleagues in the US showed that rat cells exposed to normal body levels of melatonin produced proteins important for bone formation. Because melatonin levels decrease with age, it is possible that melatonin loss is implicated in osteoporosis. Does this mean that exposure to EMR contributes to osteoporosis by lowering melatonin levels? While there are no answers to this at present, the possibility alone is yet another inducement to avoid unnecessary exposure. (Roth et al. 1999.)

Not only does it reduce melatonin, but EMR contributes to cancer by inhibiting the effectiveness of anti-cancer drugs. Several researchers have found EMR counteracts the effects of the drug Tamoxifen which is used to reduce the proliferation of cancer cells.

Quite apart from its effect on melatonin levels, EMR from powerline sources has been shown to contribute to cancers by accelerating the growth of existing tumours in the body. This occurs because cancer is much more conductive than normal tissue. (Joines et al. 1980.)

A summary of studies which have found that EMR affects melatonin levels and the effectiveness of Tamoxifen can be found on pages 218–21.

A dangerous cocktail

EMR has been shown to have a synergistic effect with chemicals. In other words, exposure to EMR *and* a chemical has a more profound effect than exposure to just one or the other. These effects have been found to exist for in vitro (test tube) studies, in vivo (animal) studies, and in studies of human populations exposed to a cocktail of emissions from a range of sources.

Drs. Meike Mevissen and Wolfgang Loscher of Germany investigated the effects of EMR and chemicals over a number of years. In a series of studies, the researchers exposed rats to a chemical carcinogen and to 50 Hz magnetic fields separately and together. They found that the combination of chemical and magnetic field resulted in larger and more frequent mammary tumours. According to the researchers, 'The data add further evidence to the hypothesis that hormone-dependent tissues such as breast might be particularly sensitive to MF- [magnetic field-] effects and indicate that immune system depression is involved in the increased breast cancer growth observed in MF exposed rats.' (Mevissen 1996.)

In 1999 a study published on the EMR/chemical combination aroused controversy. Dr. Ross Adey found that rats exposed to a carcinogenic chemical and EMR from a mobile phone had slightly *fewer* tumours than their less-exposed comrades. This has sometimes been argued to suggest that EMR has a protective effect. However, a more sinister interpretation is also possible. Adey has suggested that EMR may be killing the cells so that they are not able to proliferate.

Two studies have investigated the effects of chemicals and EMR on children. In 1998 Maria Feychting examined data on 127,000 children living within 300 metres of transmission lines in Sweden. She found that children living in areas of greater pollution from motor-vehicle exhausts had an increased risk of cancer. (Fechting et al. 1998.) The following year H. Wachtell found that children exposed to traffic exhaust and living in homes with wiring configurations that suggested a high flow of current had a higher incidence of cancer. (Wachtell et al. 1999.)

More detailed information about studies that have found a synergistic effect between chemicals and EMR can be found on pages 221–2.

Effects on cells

A large number of studies has investigated the effects of EMR on cells. These are significant because they show a broad spectrum of results likely to be applicable to living creatures in general.

One explanation for the way that EMR impacts on the body is that microwaves act on the membrane of the cells. Specifically, they are thought to interact with the proteins that are encased in the cell's membrane, causing changes to the behaviour of the cell, itself. However, it is also possible that EMR acts on electrons moving within the DNA of the cell. (Blank and Goodman 1997.) More recently, Kirschvink suggested that EMR may be absorbed in the cell by tiny particles of magnetite that he suggests are present in most tissue. (Kirschvink 1996.)

Many cellular studies have focused on systems known to be associated with cancer. Several studies have found that EMR affects the proliferation of cells; others that it stimulates the production of heat-shock proteins that are important for controlling stress in the cell.

EMR has also been shown to affect signal transduction processes by which cells receive messages from each other and the environment. This occurs through the activity of enzymes (that group of proteins which signals the body to initiate activity and are dependent on calcium) and neurotransmitters (messengers).

Exposure to modulated EMR stimulates the release of the enzyme ornithine decarboxylase (ODC). ODC is present during the development of cancer, and is generally regarded by biomedical researchers as an indicator that cancerous changes are taking place. Another category of enzymes affected by EMR is the protein kinase group. When the protein tyrosine kinase is activated by EMR, it begins a cascade of events that leads, in some cases, to unchecked proliferation of cells, which is a hallmark of cancer.

EMR has also been shown to affect the behaviour of calcium ions which are the main messengers of the body both between and within cells. In 1975 Susan Bawin and W. Ross Adey reported that exposing nerve cells to a 16 Hz field caused calcium ions ($Ca++$) to flow from the cells. (Bawin et al. 1975.) This effect, known as calcium ion efflux, has since been demonstrated by several other researchers at slightly different frequencies. Changes to these ions are consistent with abnormal growth associated with cancer.

Calcium is the most abundant mineral in the body, and is necessary for the normal function of nerves and muscles. It is involved in 'muscle contraction, bone formation, cell attachment, hormone release, synaptic transmission, maintaining membrane potentials, function of ion channels, and cellular regulation. It also serves as a second messenger in neural function in which the concentration of calcium inside the cell regulates a series of enzymatic events caused by kinases'. (NRC 1996.) Combined with phosphorus, it forms calciumphosphate in the bones and teeth. Moreover, calcium ions are important for regulating the pineal hormone melatonin which, as we have already seen, is important for protecting against cancer-causing free radicals. Calcium ion efflux would reduce melatonin levels and increase the risk of cancer.

Exposure to EMR activates glutamate, which is an important neurotransmitter in the nervous system and affects brain-wave patterns. Problems with glutamate metabolism can lead to Parkinson's disease, so it is not surprising that there is some evidence that EMR exposure could be implicated with this condition.

Because signal transduction processes such as these help to regulate gene expression, metabolism, proliferation, and differentiation of cells, interference by EMR could be enormously serious for health.

Dr. Peter French and his team have shown that cells exposed to 835 MHz—in the mobile-phone range—for 20 minutes, three times a day, for seven days, exhibited changes in growth, shape, and secretion. One of cell types studied, the mast cell, is involved in asthma, suggesting that EMR may exacerbate this condition. (Donnellan et al. 1997.) Given that the rate of asthma has increased by 50 per cent in the last 25 years and that it now affects around 150 million people worldwide, most of them in developed countries, this connection may be a worthy area of future research.

Mast cells

EMR may be activating mast cells to produce a range of other allergic effects. Drs. Shabnam Gangi and Olle Johansson at the Karolinska Institute in Stockholm have developed a compelling hypothesis that mast cells are implicated in the allergic responses that are often experienced by people after prolonged exposure to EMR, especially from computers. This is based on a number of observations. First, people with such symptoms have a higher number of mast cells, larger mast cells, and a different pattern of mast cell distribution. Second, mast cells release histamine, which causes symptoms such as itching and erythema, that are often reported by people exposed to EMR. Third, EMR is known to affect mast cells, as Dr. Peter French and his collaborators have shown.

To explain the process by which this occurs, the researchers suggest that EMR can activate a cell by affecting a neuropeptide (a chain of amino acids), or by changing the concentration of ions in the cell. Once activated, the mast cell releases its contents which include histamine. Histamine interacts with the surface of other cells, causing a range of effects such as the contraction of muscles and airways, the dilation of veins, and the secretion of gastric acid, or by inhibiting the immune system.

Not only are mast cells and the histamine they contain likely to be responsible for asthma and other allergic reactions, they may be implicated in a range of health problems that have been associated with EMR. The heart, which is known to be affected by EMR (which causes changes to ECG patterns, heart rate, and blood pressure) contains mast cells. Histamine is also contained in some cells in the endocrine system (which controls our hormones), the central nervous system, and the immune system. All of these have been shown to be impacted by EMR, as you have read in chapter five. Histamine is also found in the blood, lymphatic vessels, and the respiratory tract. (Gangi and Johansson 2000.)

Given their susceptibility to radiation and their location in many systems of the body, mast cells may thus feature prominently along the route to EMR-induced health problems.

Effects on genes

One of the most startling findings to emerge from research is that exposure to EMR can cause breaks in the genetic material of which our bodies are made. In 1995 US researchers Henry Lai and Narendra Singh created a stir in the scientific community when they showed that exposing rats to frequencies of 2450 MHz (the frequency at which microwave ovens operate) for two hours produced breaks in the DNA of rats' brains. (Lai et al. 1995.)

The following year, they showed that the same frequencies produced both single- and double-strand DNA breaks. (Lai et al. 1996.) Then, in 1997, the team showed single- and double-strand DNA breaks could also be caused by exposing rats for two hours to a 60 Hz magnetic field of 5 mG and under. (Lai et al. 1997.)

This work is enormously important. Each single cell in the human body comprises DNA, made of semi-conductive proteins, in the shape of a coiled ladder with three billion rungs. Lai's work shows that breaks can occur on one side of the ladder (single-strand breaks) or on both sides (double-strand breaks). If double-strand breaks occur, the cell will probably die and be replaced (except in the case of brain cells). If single-strand breaks occur, the body's repair system jumps into action. However, problems are most likely to occur if there are multiple single-strand breaks close together. In this case, the body may not repair the breaks in the correct order, causing the cell to survive as a mutation. Because nerve cells are poorly equipped to repair DNA damage, such breaks may impair nerve cells more than other types of cells.

This could, of course, have enormous ramifications for human health because, as Lai and Singh write, 'DNA strand breaks may affect cellular functions, lead to carcinogenesis and cell death, and be related to onset of neurodegenerative diseases.' (Lai and Singh 1997.) This could, perhaps, explain why diseases such as Alzheimer's and Parkinson's have been found among people with high EMR exposure.

Lai and Singh are not the only researchers to have found that EMR affects genes. Dr. Peter French's experiments on mast cells showed that EMR exposure changed the DNA of cells by turning some genes 'on' and others 'off'. (Harvey 2000.) Other researchers to have found genetic effects include Drs. Phillips, Adey, and Somar-Sarkar.

Several studies have found that exposure to EMR increases the expression of a number of genes associated with cancer, some of which are important for controlling the disease. Some studies have found that exposure results in changes to the chromosome envelopes which contain the genes. Others have found changes in patterns of DNA within the brain.

How significant these changes are is difficult to say. In some cases, it may be generations before we understand the full ramifications of the effects.

No, George. From my own experience and from the research I've been looking at, I can say with absolute certainty that mobile phones do not cause any genetic effects.

For studies which have found that EMR affects genes, see pages 223–6.

Effects on the blood-brain barrier

Another profoundly important discovery is that EMR affects the blood-brain barrier. This barrier exists to protect the brain—and central nervous system—from incursions by harmful molecules in the blood. If breached, it could expose the brain to chemicals that interfere with its normal function.

Recently Swedish researchers Leif Salford, Arne Brun, and Bertil Persson from Lund University showed that extremely low levels of microwave radiation cause the blood-brain barrier to leak. The researchers exposed rats to two hours of radiation with SARs of 0.4 mW/kg to 3 W/Kg—all well below international standards for human exposure. They found that the exposure caused leakage of the protein albumin through the blood-brain barrier, and that greater effects occurred at the lower exposures. (Persson et al. 1997.) According to the researchers, 'Radiation from the base stations of the mobile-phone systems should therefore be enough to affect the brain. Even someone who is in the vicinity of a person making a call may be influenced by the radiation from the phone.' (*Svenska Dagbladet* 16 September 1999.) Salford's study was not a chance finding. It confirms the work of many previous researchers including Prato (1994), Rubin (1994), Salford (1994), and Schirmacher (2000).

Breaches of the blood-brain barrier have been shown to be connected with headaches (Frey 1998) and headaches are one of the most common symptoms reported by mobile-phone users. If these headaches are symptoms of a breaching of the blood-brain barrier they may be signalling that the brain is open to chemical intrusion. This could be more of a problem for those exposed routinely to chemical emissions or for people taking medication.

Heat Shock Proteins

In late 2000 Dr. Peter French described to the Senate inquiry on EMR a route he had just identified that explains how the radiation from a mobile phone can lead to cancer. 'The support for this link is based firmly on the peer reviewed and published work of other scientists internationally and supported by observations in our own laboratory.' (*Hansard* 16 November 2000.)

The first sequence in the chain of events occurs when the mobile phone deposits energy in the brain. This energy causes the production of Heat Shock Proteins (HSPs). HSPs are produced when the cell is under stress, not just from heat as the name misleadingly implies, but also from heavy metals, drugs or chemicals. Their job is to migrate to the area of the cell where protein has been affected and repair the damage.

Heavy use of a mobile phone would result in chronic production of HSPs. Chronic production of HSPs contributes to cancer in four possible ways. They increase the potential for developing tumours; they help spread cancer; they affect the development of cancer and they reduce the effectiveness of anti-cancer drugs.

This is a tremendously important finding as it helps explain what manufacturers have long claimed is impossible: how the radiation from a mobile phone can cause serious problems without significantly heating the brain.

Resonance effects

There are a number of different types of resonance effects. Perhaps the most well known is illustrated by the example of the opera singer who reaches a note which shatters a wine glass. This occurs because the wavelength of the note matches the size of the glass. In the same way, EMR can impact on an organ if both are the same size, and the same is true of the whole body, of course. This means that slightly different frequencies will affect people differently. This type of resonance may help explain some of the window effects we discussed earlier in this chapter.

Dr. Gerard Hyland believes that resonance effects can account for effects on the body at athermal levels of radiation. 'The human body is an electrochemical instrument of exquisite sensitivity whose orderly functioning and control are underpinned by oscillatory electrical processes of various kinds, each characterised by a specific frequency, some of which happen to be close to those used in GSM. Thus some endogenous biological electrical activities can be interfered with via oscillatory aspects of the incoming radiation, in much the same way as can the reception on a radio.' (Hyland 2000.)

A particular type of resonance is magnetic resonance.

It is widely accepted that EMR causes changes to the behaviour of important ions in the body. As we have already seen, EMR can cause calcium ions to flow from cells. Similar effects have also been found for sodium ions (Na+), potassium ions (K+) and lithium ions (Li+). Sodium, potassium, calcium, and lithium are fairly reactive elements because they all contain only a few electrons—out of many available positions—in their outer shell. This means that they are constantly seeking other ions to whom they can donate these electrons to attain the stability of a complete outer shell. Thus they are easy targets for interference—such as EMR.

EMR interacts with these vulnerable elements through a *magnetic resonance effect*. This occurs when a calcium ion, for instance is in a DC magnetic field *and* an alternating (oscillating) electric field at right angles to it. The effect of this combination at the intersection of the two fields is to energise the ion, changing its usual behaviour. When the alternating field is turned off the ion re-radiates the accumulated energy.

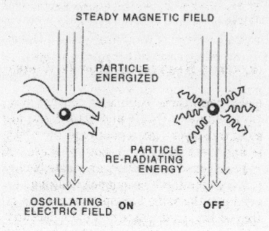

STEADY MAGNETIC FIELD

PARTICLE
ENERGIZED

PARTICLE
RE-RADIATING
ENERGY

OSCILLATING   ON   OFF
ELECTRIC FIELD

The potency of this effect is demonstrated by an experiment conducted by Drs. John Thomas, John Schrot, and Abraham Liboff in 1986. The researchers exposed a group of rats to a DC magnetic field of 200 mG, about half of the earth's magnetic field. At right angles to this field they established an alternating (oscillating) electric field of 60 Hz, such as would be emitted by a powerline or electrical appliance. The rats were positioned at the intersection of the two fields for 30 minutes. The researchers hypothesised that magnetic resonance would activate the lithium ions in the rats' brains, causing a sedating effect. The results confirmed their expectation. The rats in the resonant field showed definite behavioural changes consistent with the administration of a dose of lithium, becoming less active and more submissive. Rats exposed to either the DC field alone or the oscillating field alone showed no such changes. (Thomas et al. 1986.)

If magnetic resonance affects the ions in our bodies, the effects could be profound. Calcium is an important messenger. Sodium and potassium play a vital role in conducting electrical messages through the nerves to trigger the release of messenger chemicals that initiate action in the body. Because these ions play such pivotal roles in the nervous system, any disturbance of their function could interrupt the transfer of messages throughout the body with important consequences for health.

There are many situations in which a resonance effect such as described above could occur in our artificial and energy-dependent environment. Anywhere a conventional powerline runs parallel with a DC line such as a railway or even the earth's magnetic field, magnetic resonance has the potential to affect ions in the body. DC powerlines, which are now being constructed in several places in Australia, offer considerable opportunities for this interaction to occur—and the nervous system to be impacted.

It is also possible that fields from powerlines may resonate with radiofrequency fields to produce effects on health, though this has not been researched. If this is the case, the combination of powerlines and radiofrequency transmitters may prove to be an unhealthy cocktail in particular locations.

EMR attracts radon

Yet another explanation as to why EMR appears to cause health problems has been advocated by Professor Denis Henshaw. In 1999 Henshaw found that the radiation from high-tension powerlines may interact with radioactive pollutants to cause health problems. Placing phantom heads at various locations near powerlines, he was able to measure the amount of radioactive pollutants they accumulated. He found that heads near the lines had higher levels than controls located away from the lines, irrespective of weather conditions. From this he concluded that there is a 1.2- to 2-fold increased dose of radiation to the skin for people living under high-voltage powerlines, and that this may increase the risk of skin cancer.

In a second study, Henshaw found that powerlines ionise the air, creating corona ions, and that these ions become attached to aerosol pollutants. The result is electrically charged pollutants which are carried an average of 200 metres down-wind of powerlines, though effects were detected at 500 metres from one 275 kV line. Inhaling these charged particles could contribute to health problems, including leukemia, which has often been found among people living near power-lines. (Fews et al. 1999.)

Henshaw's theory has gained strong support from preliminary research currently being undertaken in Britain by Dr. Alan Preece. Preece, who is investigating the types of cancers among people living near high-voltage powerlines, has found that there is a significantly elevated rate of lung cancer—but only among people living down-wind from the lines.

All of the studies cited so far point to EMR impacting the body through a variety of mechanisms to cause a variety of effects ranging from inconvenient symptoms to life-threatening diseases. Even so, there are many other studies that have failed to find an association—and it is these that are usually presented as evidence that EMR is 'safe'.

However, just as a spider web does not cease to exist simply because you can't see it in the pre-dawn darkness, health effects do not cease to exist because they are not always detected.

Why studies don't find effects

One of the reasons that studies have failed to find an effect, according to EMR researcher Samuel Milham, is 'that the controls are exposed'. Reviewing research on the health risks of smoking, Milham found that a high risk was only found in studies that compared heavy smokers to non-smokers, not those comparing them to moderate smokers. In EMR research, however, it is not possible to compare high-exposure subjects to non-exposed subjects simply because there are no non-exposed people in the entire industrial world. (Milham 1998.)

Despite the fact that many laboratories have found effects from EMR exposure, many studies on animals (in vivo) and cells (in vitro) have failed to do so. It is quite likely that the outcome of laboratory studies is affected by the magnetic field of the earth at the site of the laboratory itself. In 1985 Drs. Carl Blackman and Abraham Liboff independently concluded that variations in the earth's magnetic field accounted for differences in frequencies at which calcium ion efflux was observed in different laboratories. (Becker 1990, p. 234.)

Quite apart from these difficulties, translating measurements at higher frequencies from one laboratory to another presents its challenges. According to one researcher, simply altering the position of a waste paper basket in the laboratory affects his measurements.

In many cases laboratories have been unable to confirm effects that were noted on human populations. For example, several animal studies funded by the six-year US Rapid Research program, failed to find a magnetic-field link to brain tumours or leukemias. However, this does not mean that the human studies should be ignored or that their results are invalid. Rather, it is important to understand that laboratory studies often expose their subjects to pure sine waves whereas, in real life, people are exposed to other characteristics of EMR such as harmonics, transients, and spatial and temporal changes, as we have seen in chapter two. It is possible that these confounders could be more injurious to human health than the pure undulating waves themselves.

Moreover, the RAPID program and other research conducted by the US National Institute of Environmental Health Sciences exposed animals to fields with linear polarisation for ease and thrift, rather than fields with circular polarisation. However, some studies have shown that fields with circular polarisation produce greater effects, at least on melatonin levels, and potentially, therefore, on cancer.

A fairly obvious reason why some studies fail to find effects is the difference in susceptibility of the exposed group, be they people or cells. As we have seen, children and foetuses are particularly vulnerable to the effects of EMR, as are people with extreme sensitivity to environmental conditions, and those who are genetically vulnerable.

Genetic susceptibility is a key factor in the responsiveness of an organism to EMR. A number of studies have investigated whether exposure to EMR inhibits the effectiveness of the drug Tamoxifen in reducing the proliferation of breast cancer cells. Initially, studies in three laboratories, which purchased breast cancer cells from different sources, found different results. When the laboratories used cells from a common source, they achieved consistent evidence of effects. In September 2000 Dr. Ted Litovitz related to a Senate Inquiry on EMR his own research experiences. Litovitz initially found a twofold increase in abnormalities among chicks exposed to EMR but later found no effect whatsoever. What varied in these experiments was the strain of hens used. 'We found an enormous genetic compound in the response of chick embryos to electromagnetic fields.'

Litovitz also reported his research indicating that cells require exposure to a constant signal for ten seconds for the process of signal transduction to occur. If a signal is not constant for that time, the cell will not respond. 'This is biologically enormously sensible because the cell has to be an efficient producer; it does not want to be turned on by every little hormone molecule that comes wandering by. It wants to respond to the average properties of its environment.'

This finding is enormously significant, as it means that studies that do not maintain a constant exposure for this vital period are fated not to show an effect.

Since this discovery, Litovitz has conducted numerous studies confirming the effect. He has repeatedly shown that EMR elicits an effect only if exposure is consistent for ten seconds and not if it is flashed on and off every few seconds. Further, he has found that cells respond in a similar fashion whether they are exposed to ELF from powerlines or radiofrequency from mobile phones and towers. 'The cell's characteristic response to a mobile phone is the same as that to a powerline.' (*Hansard* 22 September 2000.)

A final, and rather cynical explanation as to why EMR has not always been found to damage health is that it would be monumentally inconvenient for particular interests to do so. Imagine the consequences for industry if telecommunications technology—from mobile phones to towers and VDUs—were found to be dangerous to health. Profitable networks could crumble, financial empires could topple, litigants could demand compensation for years to come. Imagine the consequences for government if EMR were proven dangerous. Not only could it lose a huge source of revenue, but it could be faced with hugely expensive compensation claims.

But is it fair to suggest that industry bias could actually affect the outcome of scientific research? Without a doubt. Richard Smith, editor of the British Medical Journal, conducted a review of nearly 200 medical articles. He found that the vast majority which found in favour of a particular drug or tobacco were those funded by the industries under investigation. 'Almost all funding comes with strings attached,' said US Professor Hurst Hannom. 'At a minimum the recipient must be accountable for how grants are spent. At a maximum, the recipient must deliver a particular product that is acceptable to the donor.' (*Reuters* 31 July 1998.)

Similarly, an editorial in the *Lancet* of April 2001 contained a damning indictment of drug companies and the way in which they have manipulated research that might equally be applied to the telecommunications industry: 'Efforts by drug companies to suppress, spin, and obfuscate findings that do not suit their commercial purposes were first revealed to their full, lethal extent during the thalidomide tragedy ... The issue at stake ... is the relation between a company that is sponsoring a study in some way and the investigators ... In addition, the sponsor's role in interpreting data, writing the report, or publishing the paper is far from clear, leaving a damaging ambiguity over the entire research process.' (*Lancet* editorial, 14 April 2001.)

The situation is no different in Australia, according to a report on academic freedom released by the Australia Institute in March 2001. After conducting a poll of academics, it found that '17 per cent of respondents reported that they had experienced being prevented from publishing contentious results ... Forty-one per cent reported that they had experienced discomfort with publishing contentious research results ... and almost half (49 per cent) reported that they had experienced a reluctance to criticise institutions that provide large research grants or other forms of support'. (Kayrooz et al. 2001.)

So who has funded research on EMR?

In the western world, the majority of research on this issue has been conducted by the power and telecommunications industries. Some of the major research projects have been conducted in the US by the Wireless Technology Research (WTR) project of the 1990s, funded to the tune of $27 million by telecommunications companies; and the State Powerlines Project of the 1980s, funded by the Electric Power Institute. Additional studies have also been funded by individual industries, notably Motorola.

And industry has not always played fair. Leading Australian science and technology writer, Stewart Fist has stated, 'I have no reservations whatsoever in claiming the cell phone industry has manipulated scientific evidence in the same blatant way the tobacco industry did for many years.' (*The Australian* 1 June 1999.) Stewart has compiled a thorough expose of industry manipulation of mobile-phone research during the 1990s on his website. (http://www.electric words.com/junk/carlo/carlo.html)

Several examples of this manipulation are worth noting. In 1995 Henry Lai and NP Singh found that exposing rats to the radiation from mobile phones resulted in breaks in the DNA of their brains. This research came under strong fire from the mobile-phone industry for some time until Motorola's Norm Sandler wrote in a memo, 'I think we've sufficiently war-gamed the Lai-Singh issue.' War-gamed! As if this were not enough, Motorola later asked Lai to change the report of a new study about DNA breaks. In 1996 that same company attempted to restrict Dr. Ross Adey from discussing his finding that mobile-phones radiation affected brain tumours in rats, even though he had found that it *inhibited* the number of tumours.

One of the ways in which industry and governments have subtly manipulated research results is their failure to fund studies which are likely to prove that EMR does produce adverse effects—or researchers who have spoken publicly about the risks. Casualties of this sort are US scientists Henry Lai and Ross Adey, who discussed their funding difficulties on ABC TV's *Panorama* on 'Four Corners' (5-6 July 1999). Closer to home, Sydney's Dr. Peter French has repeatedly failed to

accurately portraying the risks. Those of us who intend to speak for public health must be ready for opposition that is nominally but not truly, scientific.' (Goldsmith 1995.)

Until governments insist that research be conducted by genuinely independent scientists, funded by telecommunications companies but allocated by independent panels with representation from a variety of bodies, science will continue to be a political game, played for economic profit rather than the pursuit of knowledge and for the benefit of humanity.

-7-

Just in case

I f you wear a seatbelt, have a burglar alarm in you home, use natural products, buy food without artificial colours or flavours or lock you car when you get out, you already take precautions. We take precautions every time we cross a road, light the gas fire, sterilise the baby's bottle or have a medical check up. Taking precautions is not about believing the worst is nigh. It's about being sensible; it's about minimising risks where you can.

There are good reasons to take precautions to protect ourselves from exposure to EMR. There is now a wealth of scientific and community experience showing that EMR from power and telecommunications systems affects our bodies. Just how potent are these effects, and what the long-term results of this exposure are likely to be, nobody yet knows. Yet with the memory of the asbestos and tobacco debacles (to name but two) fresh in society's collective mind, it is clear that, in the long term, it is just not worth taking risks.

Taking precautions is not only a sensible way of protecting our health, it makes good economic sense. Should science conclusively prove that EMR causes health problems, the gates will be opened for a flood of litigation, with affected people claiming damages and compensation from governments, utilities, employers, and mobile-phone manufacturers. Already the writ is on the wall, so to speak, with US superlawyer Peter Angelos, champion of the people against the tobacco and asbestos industries, signing up to assist the attorney for brain tumour victim and mobile-phone user Dr. Christopher Newman.

There are also social reasons why we should take precautions; considerations about the degree of risk society is prepared to take. Swiss insurance company, Re, addressed this issue in a report on the insurance implications of EMR exposure. According to the document, the risk of legal liability from EMR exposure 'is not incalculably small, but in view of the conceivable extent of damage, incalculably great'. It argues that 'the sociopolitical risk ... must be classified as extraordinarily high', irrespective of actual health effects, as society's attitudes towards acceptable risks change. (Swiss Re 1996.)

Certainly industry is acutely aware of the risks of EMR. Despite its claims that EMR is not a health risk, it has been taking rear-guard action, just in case it is. In 1997 it was reported that British electrical utilities were establishing an eight million pound legal defence fund in anticipation of future litigation for health claims. (*Daily Mail* and *Guardian* 12 October 1996.) And at least five manufacturers of mobile phones have taken out patents for low-radiation devices, some as early as six years ago. An application by Hitachi claims its devices prevent 'the health of the user from being injured', and other applications refer to 'health risks', 'safe distances', and 'radiating systems'. (*UK Independent* 25 October 1998.)

You blithering idiot, Fothersgill! Just because I said mobile phone base stations are perfectly safe on the news last night, doesn't mean you can put one up outside my place this morning!

Whatever the motivation, the outcome is the same. Taking precautions is, in the long run, the only logical response to a situation that involves risk. Taking precautions benefits the community, industry, and governments alike.

Precautions for power sources

Prudent avoidance

'Prudent avoidance' is a concept usually applied to reducing exposure to EMR from power sources. It was first coined by Drs. Morgan, Florig, and Nair from Carnegie Mellon University in their 1989 report to the US Office of Technology Assessment, an advisory body to the US President. In this document, the authors defined prudent avoidance as 'taking steps to keep people out of fields by rerouting facilities and redesigning electrical systems and appliances'. (WHO 2000.)

In Australia the term *prudent avoidance* was used by Sir Harry Gibbs to mean 'doing whatever can be done at modest cost and without undue inconvenience to avoid the possible risk (to health)'. In his 1991 report into high-voltage power lines, Gibbs recommended that electrical authorities adopt a preventative policy of 'prudent avoidance' regarding the selection of routes for power lines. 'The epidemiological studies into the effects, if any, of electromagnetic fields at extremely low frequencies support the view that it is possible that exposure to those electromagnetic fields causes an increased risk of developing cancer in childhood and an increased risk in adults of developing leukemia, lymphoma and brain tumours.' (Gibbs 1991.)

A policy of prudent avoidance was adopted by Australia's peak electric authority, Electricity Supply Authority of Australia (ESAA) on 19 July 1991, and was endorsed by the federal government in a document tabled in parliament in October 1996. To implement prudent avoidance, the ESAA recommends:

- 'consulting the community about the siting of new transmission lines, and routing lines away from population centres or areas where people gather. Particular attention should be paid to schools, child care centres, and other areas where children congregate; and
- designing lines and facilities so as to mitigate fields.'

(ESAA 1998.)

Overseas, there have been efforts to implement essentially similar policies of prudent avoidance. In July 1987 the Final Report of the New York State Power Line Project recommended that 'There should be a major research effort on means of power delivery and use that would reduce magnetic field exposure.'

At the same time as Gibbs was considering evidence for his report, a panel chaired by Professor Hedley Peach was compiling his own report on 'Electromagnetic Fields and Health' for the Victorian government. The panel recommended a number of prudent avoidance strategies to reduce public exposure to EMR from powersources. Among its recommendations were the following:

- 'Substations located in buildings should be designed to reduce magnetic fields within the occupied areas of the building.
- Design of new transmission and distribution facilities should be based on a policy of prudent avoidance, and consideration should be given to the proximity of houses, schools, and other sensitive locations.
- New double-circuit transmission lines should be reverse phased.
- New suburban lines should be undergrounded or bundled conductors used.
- Research be undertaken into the reduction of fields from transmission lines, home wiring systems, and household appliances.'

(Peach et al. 1992.)

In 1999, the final report of the six-year Rapid Research program run by the US National Institute of Environmental Health Sciences (NIEHS) was released, which recommended low-cost action to reduce exposure. 'NIEHS suggests that the power industry continue its current practice of siting power lines to reduce exposures and continue to explore ways to reduce the creation of magnetic fields around transmission and distribution lines without creating new hazards. We also encourage technologies that lower exposures from neighborhood distribution lines provided that they do not increase other risks, such as those from accidental electrocution or fire ... finally, the NIEHS would encourage the manufacturers of household and office appliances to consider alternatives that reduce magnetic fields at a minimal cost. We feel that the risks do not warrant major and expensive redesign of modern electrical appliances, but inexpensive modifications should be sought to reduce exposures.' (NIEHS 1999:38.)

Australia's Senate inquiry into EMR, held in 2000–1 (see pp. 118–19), was unintentionally drawn into the debate about the risks to health from power sources. Not only did it receive many submissions outlining problems that people had experienced, but Sir Richard Doll's report on the leukemia risk of powerlines was released while the committee was drafting its report. So, even though the power situation was strictly outside the committee's terms of reference, the report of the chair contained some pertinent recommendations. It recommended that:

- 'Particularly in the light of recent reports on the links between powerlines, radio towers, and leukemia, additional research into extremely low frequencies and television/radio tower exposure should be encouraged. (Recommendation 2.1.); and
- Precautionary measures for the placement of powerlines be up-graded to include wide buffer zones, and undergrounding and shielding cables where practicable. (Recommendation 2.2.)'

Prudent avoidance in action

A number of companies have introduced measures designed to reduce the exposure of their workers and, no doubt, to guard against the possibility of litigation in the future.

On 22 June 1995, library supplier RAECO signed an agreement with the Australian Services Union, stipulating that no library worker would be exposed to an average magnetic field above 4 mG—although higher exposures have been known to occur in practice. The ACTU has adopted a policy of prudent avoidance and recommended exposures of no more than 2 mG for an eight-hour day. In recent years John Lincoln has worked with a number of companies in Sydney to help reduce the exposure of their employees. It is encouraging to see that, as awareness about EMR risks grow, the demand for this service increases.

In Sweden, where an unusually high number of people suffer from allergies to electricity—most often as a result of prolonged computer use—some admirable efforts to protect people from EMR have been implemented. The Swedish government has adopted a 2 mG limit as a guideline in its official policy on prudent avoidance in planning. This means that new children's facilities can be built only in areas where the ambient EMR exposures are less than 2 or 3 mG. It also obliges builders and engineers to construct or renovate buildings in such a way that exposures are reduced to 2 or 3 mG. (*Low-frequency Electric and Magnetic Fields—the Precautionary Principle for National Authorities— Guidance for Decision-makers* 1996.)

The Swedish Union of Clerical and Technical Employees in Industry (SIF), has developed a policy of 'Noll Risk' (No Risk) in the IT environment, and aims to reduce workers' exposure and, as a result, exposures in the external environ- ment. It aims to provide information so that consumers can chose low-emission products and industry can implement appropriate precautions to protect its workers. (Swedish Union 1999.)

Low-emission computer monitors, now in widespread use throughout the world, originated in Sweden. Negotiations between consumers, trade unions, and health and safety bodies led to the establishment of a new emission level of 2.5 mG from computer screens. (*MPR*3 1995.)

The Swedish telecommunications company Ellemtel is one company that has taken action to accommodate the problem of a large number of employees allergic to electricity. By 1990 the company had 39 employees with such allergies whose conditions related to long hours of computer use. In order to retain this valuable bank of expertise, Ellemtel expended extensive resources on providing a work environment with reduced electromagnetic fields. This included changing and shielding computer monitors, grounding desk lamps, adjusting wiring, rearranging office furniture, turning off equipment when it was not in use, reducing chemical exposure, and shielding the transformer and introducing a buffer zone around it. Especially shielded rooms—with a field of less than 0.2 mG—were provided for allergic employees. Following these measures, the company reported a decrease in the number and severity of new cases of electrical allergy.

Because people with allergies to electricity experience severe reactions in hos- pital, there was a strong demand in Sweden for medical care that accommodated their sensitivity. The county of Norrbotten in the north of Sweden responded to the challenge by constructing a hospital designed to reduce EMR exposures. As well as locating transformers in a separate building, designers placed wiring in shielded plastic tubes, and shielded controls and circuit breakers. Special rooms for patients with allergies to electricity have been coated with aluminium. (Network News, Summer 1997.)

Germany was the first country in the world to introduce a national standard for EMR exposures from power sources. In January, 1997, Ordinance 26 of the Federal Emission Control Act came into force, requiring that people living in residences, schools, hospitals, day-care centres, and playgrounds were exposed to no more than 1 G (1,000 mG) and 5 kV/m from new or upgraded electrical equipment, though higher exposures were permitted for short periods in some outdoor areas. Existing facilities were given three years to comply with the new limits. While the concept of limiting exposures in sensitive areas is commendable, these fields are, in my opinion, still far too high to adequately protect people's health.

Britain has also implemented strategies to protect the public. Already several companies have commissioned buildings to reduce EMR exposure to workers to less than 3 mG away from office equipment. (*EMRAA News* June 1998:3:2.)

In 1999 the UK Local Government Association, which represents 50 million people, issued a strong statement on the need to adopt precautions to protect community health from EMR. The document commented that 'it is surely unacceptable that the public's health should be subjected to the possibility of compromise through the application of arguments which appear to be based on the need to prove, beyond all reasonable doubt, a causal link in terms of absolute scientific proof'. It recommended that restrictions be placed on new developments near powerlines, and called for further research into the health effects of EMR. (*EMRAA News* June 1999:4:2.)

In Italy the environment ministry released a draft report which included precautionary suggestions to protect children. The ordinance recommended that magnetic field exposures be kept to a maximim of 2 mG in new schools, kindergartens, and playgrounds built next to powerlines. (*MW News* March/April 2000.)

Precautions for telecommunications sources

In 1996, the Department of Communications commissioned a survey by AGB McNair on community concerns about the health effects of EMR. It found that 40 per cent of mobile-phone users were concerned about the health effects of the devices. Further, 44 per cent of general respondents and 50 per cent of people living near the facilities were concerned about the health effects of mobile-phone base stations. (McNair 1996.) No doubt, in light of the increasing proliferation of this technology and the release of more scientific evidence of risk in recent years, community concern is likely to have grown.

The precautionary principle

Similar to the policy of prudent avoidance, the precautionary principle is an approach to reducing risks that is usually applied to telecommunications equipment and infrastructure. It is based on the position that it is not necessary to have definitive scientific proof that a human activity (such as installing EMR-emitting technology) causes a particular effect (such as health problems) in order to justify taking precautions.

The Precautionary Principle was recognised at the 1992 UN Conference on Environment and Development, and included Principle 15 of the Rio Declaration which reads: 'In order to protect the environment, the precautionary approach should be widely applied by States according to their capabilities. Where there are threats of serious or irreversible damage, lack of full scientific uncertainty shall not be used as a reason for postponing cost-effective measures to prevent environmental degradation'. This principle was later included in Australia's Inter-governmental Agreement on the Environment (p. 25).

Around the world a number of policies have been designed which have sought to reflect a precautionary approach to using mobile phones and siting base stations.

It's the Emperor's idea. Something about taking precautions.

Vienna EMF resolution

In October 1998 a symposium held in Vienna on 'Mobile Phones and Health' discussed the possible biological and health effects of RF electromagnetic fields. Many well-known scientists in attendance who had been working in the field for many years endorsed the following resolution:

Preamble: The participants agreed that biological effects from low-intensity exposures are scientifically established. However, the current state of scientific consensus is inadequate to derive reliable exposure standards. The existing evidence demands an increase in the research efforts on the possible health impact and on an adequate exposure and dose assessment.

Base stations: How could satisfactory Public Participation be ensured? The public should be given timely participation in the process. This should include information on technical and exposure data as well as information on the status of the health debate. Public participation in the decision (limits, siting, etc.) should be enabled.

Cellular phones: How could the situation of the users be improved? Technical data should be made available to the users to allow comparison with respect to EMF-exposure. In order to promote prudent usage, sufficient information on the health debate should be provided. This procedure should offer opportunities for the users to manage reduction in EMF-exposure. In addition, this process could stimulate further developments of low-intensity emission devices.

The resolution was signed by Carl Blackman, Neil Cherry, G. Kas, Lebrecht von Klitzing, Wolfgang Kromp, Michael Kundi, Henry Lai, William Leiss, Theodore Litovitz, Kjell Hansson Mild, Wilhelm Mosgoller, Joachim Roschke, Felix Schinner, Stanislaw Szmiegielski, Luc Verschaeve, and Ulrich Warnke.

The Salzburg resolution

In June 2000, a number of leading scientists endorsed precautionary recommendations on the siting of mobile-phone base stations, known as the Salzburg Resolution. This document states:

1. It is recommended that development rights for the erection and for operation of a base station should be subject to a permission procedure. The protocol should include the following aspects:
   - prior notification and active involvement of the local public
   - inspection of alternative locations for the siting
   - protection of health and wellbeing
   - considerations about conservation of land- and townscape
   - computation and measurement of exposure
   - considerations about existing sources of HF-EMF exposure
   - inspection and monitoring after installation.
2. It is recommended that a national database be set up on a governmental level giving details of all base stations and their emissions.
3. It is recommended for existing and new base stations to exploit all technical

possibilities to ensure exposure is as low as achievable (ALARA-principle) and that new base stations are planned to guarantee that the exposure at places where people spend longer periods of time is as low as possible, but within the strict public health guidelines.

4. Presently the assessment of biological effects of exposures from base stations in low-dose range is difficult but indispensable for protection of public health. There is at present evidence of no threshold for adverse health effects. Recommendations of specific exposure limits are prone to considerable uncertainties and should be considered preliminary. For the total of all high frequency irradiation a limit value of 100 mW/m² (10 µW/cm²) is recommended. For preventive public health protection a preliminary guideline level for the sum total of all immissions from ELF pulse modulated high-frequency exposure facilities such as GSM base stations of 1 mW/m² (0.1 µW/cm²) is recommended.

(Salzburg Resolution on Mobile Telecommunication Base Stations International Conference on Cell Tower Siting Linking Science & Public Health, Salzburg, 7–8 June 2000.)

The Stewart report

In May 2000, after months of investigation, an independent panel of 12 eminent biological research scientists, chaired by Sir William Stewart, released a report (dubbed the Stewart report), which recommended precautions to the siting and use of mobile phones and associated infrastructure. According to the report, 'if science has greater power to do good, it also has greater power to do harm. They [the committee] therefore advocate a precautionary approach to new technology where there are uncertainties about the associated risks'.

Amongst its recommendations, the report suggested that:

- exposures be reduced from the levels of the NRPB guidelines to those of the ICNIRP guidelines (note that the levels recommended by ICNIRP allow higher exposure than those presently used in Australia);
- carriers require council planning permission for all base stations, including those under 15metres which are now exempt;
- councils should be notified of *all* proposed base stations;
- RF emissions of all base stations should be kept to lowest practical levels
- the beam of greatest RF intensity from a base station inside or outside a school yard should not fall on any part of the school property without permission from parents and the school;
- warning signs should be erected at entrances to hospital buildings indicating that mobile phones should be turned off;
- a national data base providing details of all base stations and their emissions

should be constructed;

- An independent, random, and ongoing audit of all base stations should be made to ensure standards or specifications are not exceeded.

The report took a particularly strong stand on the risks associated with mobile-phone use by children. It suggested that, because children may be 'more vulnerable because of their developing nervous system, the greater absorption of energy in the tissues of the head ... and a longer lifetime of exposure', mobile phones not be used by children under 16, and that manufacturers refrain from advertising mobile phones to children. Sir William said that, while he used a mobile phone himself, he would not want his grandchildren to use one.

It is worth noting that, in arriving at its precautionary conclusions, the committee was dealing with data that could be considered to be in some ways biased. Firstly, it gave full consideration to studies that had been conducted by industry; secondly, it was denied vital information. When it requested copies of research including a study on the effects of shortwave radar on children in Latvia, the National Radiological Protection Board replied that the study had not been published and was not available. However, this is not the case, as the study was published in 1996 and is easily accessible by the public. Had this bias been taken into account, there would have been even more grounds for precaution.

The Australian Senate's inquiry into EMR

During 2000 and the early part of 2001 a Senate inquiry was held in Australia to investigate the health effects of EMR and the adequacy of research and standards-setting procedures. The process was characterised by jockeying of party-political interest groups and included, as well as reports of some new research discoveries, some enlivened debate. Commenting on the inquiry, the chair of the committee, Democrat senator Allison said, 'I became aware that there has been pressure from industry on participating senators to discredit some witnesses.' (*Address to the RF Spectrum Conference*, Sydney, 23 March 2001.) As a result, the committee failed to reach consensus and three reports were handed down: one from the chair and two dissenting reports from the Labor and Liberal senators respectively.

The report of the chair contained some sensible precautionary recommendations including:

- 'that the Commonwealth Government considers developing material to advise parents and children of the potential risks associated with mobile phone use.' (Recommendation 2.3.)
- 'shielding and hands-free devices are tested, labelled for their effectiveness

and regulated by standards.' (Recommendation 2.4.)

- 'the development of an industry code of practice for handling consumer health complaints.' (Recommendation 2.6.)
- 'the establishment of a centralised complaints mechanism in ARPANSA or the Department of Health for people to report adverse health effects associated with mobile phone use and other radiofrequency technology, and for the data from this register to be considered by the NHMRC [National Health and Medical Research Council] in determining research funding priorities.' (Recommendation 2.7.)
- 'that the equivalent of $5 for each mobile phone in use be collected annually for this purpose [independent research] (approximately $40 million).' (Recommendation 3.1.)
- 'that funding for maintaining the NHMRC-administered research program be provided at, say $4 million per annum of the $40 million and that the balance be used by the CSIRO to establish a structured program of research and set up a specialised research unit for this purpose.' (Recommendation 3.2.)
- 'that the radiofrequency standard be defined and administered by a process similar to that used by Standards Australia.' (Recommendation 4.1.)
- 'that the level of 200 microwatts per square centimetre in the expired Interim Standard (AS/NZS 2772.1(Int):1998) be retained in the Australian Standard.' (Recommendation 4.2.)

(*Inquiry into Electromagnetic Radiation* May 2001.)

The precautionary approach in action

Mobile phones

In March 2000 Sydney's Royal North Shore Hospital made headlines around the country when it introduced a landmark directive on mobile-phone use. It advised the use of a landline phone or pager in preference to a mobile phone, and suggested that staff who choose to use mobiles hold them at a distance of three to four centimetres from the head. It further suggested that mobile phones be used in the open or near a window to obtain a clear signal without having to increase power, and advised staff not to use a mobile phone when driving unless they had a hands-free facility in a car fitted with an external aerial.

In Britain, the Public and Commercial Services Union has taken the precedent-setting stand of advising its 266,000 members to stop using mobile phones so as to protect their health. Espousing the message, 'Don't gamble with your health', the union stated that members must not be forced to carry or use a mobile phone. It also recommended that:

- 'phone charge cards be provided to staff;
- workers requiring a mobile while travelling leave the devices turned off for the majority of the time;
- incoming calls be acknowledged and returned later from a land line;
- mobiles not be carried next to the body when operational;
- mobiles be kept centimetres away from the head during calls. (*Sunday Mirror* 19 March 2000.)'

In July 2000 UK education secretary David Blunkett sent guidelines to every school in England to discourage the use of mobile phones by children under 16. His letter stated, 'Children aged 15 and under … are likely to be more vulnerable to any unrecognised health risks from mobile-phone use than are adults because their nervous systems are still developing. Also, because of their smaller heads, thinner skulls and higher tissue conductivity, children may absorb more energy from a mobile phone than do adults.' It also recommended that, 'Where children do use mobile phones, they should do so for as short a time as possible.'

In December 2000 the UK's Department of Health released two leaflets containing precautionary recommendations which were sent to stores to be distributed with pre-Christmas purchases of mobile phones. The first, entitled 'Mobile Phones and Health', suggested ways of limiting exposure, in line with the recommendations of the Stewart report:

'If you use a mobile phone you can choose to minimise your exposure to radio waves. These are ways to do so:

•keep your calls short

•consider relative SAR values … when buying a new phone.'

The leaflet advised against the use of mobile phones, whether or not they are hands-free, while driving, and suggested 'widespread use of mobile phones by children (under the age of 16) should be discouraged for non-essential calls … In the light of this recommendation the UK Chief Medical Officers strongly advise that where children and young people do use mobile phones, they should be encouraged to:

- use mobile phones for essential purposes only
- keep all calls short—talking for long periods prolongs exposure and should be discouraged.'

Even before this initiative was announced, one local council in Britain issued guidelines advising children how to reduce health risks. Brighton and Hove Council in East Sussex advised the 55,000 students in its constituency to use a landline in preference to a mobile phone; to hold mobile phones away from the head during calls; to keep calls short; and to choose a mobile phone that uses lower frequencies. Following this example, Edinburgh council began drafting guidelines for mobile-phone use for the 60,000 students in its jurisdiction. (*Newsquest Media* 17 January 2000.)

In mid-1999, London's Metropolitan Police Force issued guidelines for those among its 27,000 members concerned about the safety of mobile phones. The guidelines stated: 'Does the user really need to use a mobile phone? If so, they should limit the length of time to certainly no more than five minutes. If users are required to make regular and lengthy use of mobile phones, there would be no harm in using an earpiece.' (*BBC News* 2 June 1999.)

The UK Defence Evaluation and Research Agency (DERA), despite its public skepticism about EMR from mobile phones being harmful, nevertheless adopted precautions to protect its employees from risk. The agency ordered at least two batches each of 20 protective Microshield mobile-phone pouches. (*EMRAA News*, September 1999.)

On 8 December 2000 the German Academy of Pediatrics issued a statement advising parents to restrict their children's use of mobile phones. 'Unnecessary, frequent and extended use are to be strongly discouraged. Children only need mobile phones to communicate very infrequently, in exceptional situations.' While the statement encouraged all mobile-phone users to keep conversations brief, it recommended particular precautions be applied to children by virtue of the risk posed by their growing bodies. (*MW News* January/February 2001.)

In the US, the Disney Corporation also responded to concerns about the safety of radiation from mobile phones. Just before Christmas 2000, the company announced that it would withdraw all commercial links to the mobile-phone industry, with the result that its popular cartoon characters would not appear on mobile phones until science establishes the radiation is safe for children.

There have also been some educational initiatives to limit the use of mobile phones by children, although not always in the interest of protecting their health. In Utah, education officials banned the use of pagers, mobile phones, and other 'interfering devices' that disrupt learning in the classroom. (*Utah News* 19 October 1999.) In Sydney, Blakehurst High School banned mobile phones from school as a result of concerns about theft, and many other schools in the Sutherland and St George areas now require mobile phones to be turned off when at school. (*St George and Sutherland Shire Leader* 27 March 2001.)

Mobile-phone base stations

Adopting a precautionary approach to the siting of mobile-phone base stations may be essential if councils, property owners, and industry are to avoid future compensation claims. To the question 'who will be at risk from legal claims?' UK lawyer Mr Alan Meyer replies,

'The mobile phone network operators who usually own and erect their own, or share masts for the antennae and dishes.

'The landowner, private or public, who leases or sells the land to enable the masts and base stations to be erected by the network operator in exchange for an annual rental or sometimes a sale price.

'The Authority, who perhaps sanctioned the Landowner to allow the Mast to be erected on its land or buildings ie, Schools, Police and Fire Brigade Authorities together with Health Authorities responsible for Hospital Land and Buildings.

'The Insurers of all such people and bodies.'

(Alan Meyer, 'Mobile Phones and Mobile Phone Networks Potential Litigation or Law Suits', presented at Gothenburg conference on Mobile Phones, September, 1999, available on the EMRAA web page.)

One company to take the risk of compensation claims seriously is Australian insurer, Mercantile Mutual. When, in 1997, Vodafone workmen arrived to install antennas on the insurer's building, company officials refused them entrance. According to a company spokesman, 'The risk to health … may expose us as owners of the property to liability for injury to persons who are or who are alleged to have been exposed to emissions from the base station. The amount of such claims is impossible to calculate.' (*The Australian* 4 July 1997.)

Along with its precautionary leaflet on mobile phones, the UK's Department of Health released a second leaflet, called 'Mobile Phone Base Stations and Health', which dealt with a variety of issues pertaining to networks and the distribution of emissions from an antenna. It said, 'Gaps in scientific knowledge led the Stewart Group to recommend a precautionary approach to the use of mobile phones and base stations until more research findings become available. They added that in some cases people's well-being may be adversely affected by insensitive siting of base stations.' The leaflet also advised that, beginning autumn 2000, the UK's Radiocommunication Agency would be conducting audits of base stations to check their compliance with the British standard.

One of the most successful and extensive examples of the precautionary approach in action in this country is the establishment of council policies on the siting of mobile-phone base stations. As described in chapter four, these policies stipulated minimum distances or exposure levels for the positioning of base stations near sensitive areas such as residences, schools, childcare centres, aged care centres, and hospitals. The first such policy was devised by Dr. Garry Smith, Principal Environmental Scientist at Sutherland Shire Council and former cancer researcher at NSW University. It recommended that mobile-phone base stations be installed no closer than 300 metres from sensitive areas unless exposure levels could be guaranteed to be less than $0.2 \ \mu W/cm^2$.

This policy was widely emulated by councils and council organisations throughout Australia. The NSW Local Government Association introduced policies recommending buffer zones extending at first 500 metres and later 300 metres from sensitive areas.

Councils are not the only bodies to recommend precautions to protect the public. In July 1996 the NSW Parents and Citizens Federation passed a resolution about the positioning of mobile-phone base stations. The meeting resolved 'that Federation calls for a stop to the installation of telecommunication mobile-phone towers in the grounds, or in the immediate vicinity of schools or preschools, due to the adverse effects to health as indicated in current research'.

Just over a year later, the NSW Minister for Education and Training stated in Parliament that there should be a 500 metre buffer zone around schools. 'The Department of School Education objects to the installation of mobile-phone towers near schools, and that normally means within a radius of 500 metres. This objection is based on a policy of prudent avoidance.' (*Hansard* 22 October 1997.)

Overseas, policies were also introduced on the siting of mobile-phone base stations. In Scotland at least half the country's councils have adopted a precautionary approach to the siting of these facilities.

In 1996 New Zealand's Christchurch Council introduced a limit on EMR to which the public can be exposed from mobile-phone base stations. The prescribed level of 2 μW/cm^2, which is 1/100th of the country's standard, has been recognised in law. (McIntyre V Bell, South New Zealand Environment Court (A96/15 NZPT 1996.) In the same year, the New Zealand Ministry of Education issued a statement prohibiting the erection of mobile-phone base stations in schools. 'Of paramount importance to the ministry is the provision of an environment where boards of trustees, parents, teachers, pupils and other occupants of the school site can feel comfortable. For this reason the ministry has decided cell phone transmitters will not be sited on Crown-owned school sites in the future.' (*MW News* September./October 1996.)

Australia's lack of precautions

Nevertheless, Australia's telecommunications industries and relevant government departments have been somewhat reluctant to embrace the precautionary approach. The federal government has adopted the position that there is no need for precautions because 'the weight of national and international scientific opinion is that there is no substantiated evidence that living near a mobile-phone tower or using a mobile phone causes adverse health effects at typical levels in the environment'. As for the problem of people who experience allergies from exposure to

EMR, the government conveniently does not have a policy. (Correspondence 25 November 1999.)

While it is one thing to avoid taking precautions, it is quite another to launch an offensive against those who have expressed concerns about the health effects of EMR, as the Australian government has done. In 1997 the Department of Communications conducted an expensive PR campaign around the country, a focus of which was the denigration of Drs. Bruce Hocking and Neil Cherry, both of whom had shortly beforehand presented evidence of harmful effects from EMR. To add insult to injury, this project was financed from the government's $4.5 million allocation for research and public education.

Then, on 5 March 1997, communications minister Senator Alston delivered an extremely colourful tirade against visiting New Zealand physicist and university lecturer Dr. Cherry, a renowned biophysicist, meteorologist, and international speaker on EMR. Under parliamentary privilege, Alston claimed 'This bloke is a charlatan. His biographical details start off with mindless and irrelevant gobbledygook … This man is a rabid populist and totally uninterested in any considered scientific debate. His remarks are highly inflammatory … [he is] a snake oil merchant.' (*Hansard* 5 March 1997.)

Why the virulence of the attack on health-risk news-bearers?

The Federal government is a major beneficiary of the telecommunications industry. In the 1999-2000 financial year it reaped $16.6 million from carriers for annual licence fees and $1,360 million for the sale of spectrum for communication networks. In its May 2000 budget, the government announced that it was intending to balance its budget with the assistance of sales of $2.6 billion of spectrum licences. With such enormous sums at stake, it is has been less than enthusiastic to admit risks from EMR or to embrace precautions to protect public health.

This government is confident that there are absolutely no adverse health effects from electromagnetic radiation whatsoever!

If the government does not embrace a precautionary approach to the telecommunications industry, surely the relevant standard would, you might think. Sadly, however, this has not been the case. In early 1999 discussions in Standards Australia's TE7 committee on a new radiofrequency standard ground to a halt because industry refused to include a meaningful precautionary approach in the document. The committee was disbanded, and a new standards working group established.

While a precautionary approach to establishing a telecommunications network would place some restrictions on the siting of mobile-phone towers, the government has done precisely the opposite. It has introduced legislation which provides carriers with a virtual carte blanche to establish 'low impact' mobile-phone towers almost at will, by providing them with immunity from state and council legislation. These towers have been classified as 'low-impact', not because they emit less radiation, but because they are less of an eyesore. This legislation has enabled towers to be built near residences, schools, kindergartens, hospitals and other areas where they could possibly jeopardise people's health—often without affected residents being notified.

It looks OK. Might as well take it into the city and install it next to the school.

Not only is this lacking in precaution, it also runs contrary to the wishes of many people in the community.

A number of eminent researchers have also called for a precautionary approach to the siting of mobile-phone towers. Dr. Henry Lai, who found that rats exposed to EMR from mobile phones developed DNA breaks, has stated 'my opinion is that cell towers should not be installed on top of schools, day-care centers, etc., where the occupants could be more vulnerable to RFR'. (*First World Conference on Breast Cancer* 1997.)

Standards

While Australians are often told that their standards are among the strictest in the world, this is not necessarily the case. Not only is our government under pressure to adopt the more lenient standards of the ICNIRP guidelines, some countries are taking action to reduce public exposure to EMR.

On 1 February 2000 Switzerland introduced new limits that are among the strictest in the world. These limits will apply in sensitive areas such as residences, schools, hospitals, and playgrounds, but outside these areas the limits of the ICNIRP guidelines continue to apply.

The legislation requires mobile-phone base stations operating at 900 MHz to limit their emissions to 4 μW/cm^2 at any one installation. (An installation is defined as all the antennae on a particular mast or roof.) This is 100 times lower than the 450 μW/cm^2 level allowed by the ICNIRP guidelines.

The legislation also applies limits to exposures from power sources of 10 mG, which is also 100 times lower than the ICNIRP guidelines.

In Toronto, Canada, the Public Health Authority has recommended precautions to reduce exposure from mobile-phone base stations. It has suggested that public exposure be limited to 6 μW/cm^2 (or 5 V/m) for 900 MHz antennae and 10 μW/cm^2 (or 6 V/m) for 1800 MHz antennae.

Other precautionary suggestions

While technology is hurtling at breakneck speed towards faster and more diverse forms of mobile technology, the so-called third-generation systems that will allow users to access the internet from their mobile phones, for example, do so in defiance of the need for any precautions, and at huge public and corporate risk.

Until the safety of this technology is established beyond doubt, it would be infinitely wiser to utilise and design systems that expose the public to the very minimum of radiation. Some suggestions include:
- the use of CDMA technology;
- the use of hardwiring, rather than radiofrequency connections, inside buildings for phones and computers;
- the design of telecommunications networks with small cells to reduce power;
- the redesign of mobile-phone headsets to radiate power away from the head;
- the use of fibre-optic cables where possible.

Fibre-optic cables have the capacity to transmit vast amounts of information without any radiation whatsoever, without any risk to health and at far greater clarity than the existing telecommunications system. The cables convert information into beams of light which travel at breakneck speed and provide faster, cleaner data transmission in wider bandwidths. They can be used for high-quality internet, fax, telephone, radio, and television signals.

While governments profitably procrastinate, industries rake in huge annual incomes, communities cry out to be heard, individuals become sick, and scientific evidence continues to find a health link. Given that ultimate proof may, as in the tobacco debate, be years away, what is the answer? The only sensible response is one of limiting exposure.

The challenge now is for electrical authorities, councils, planners, and builders to translate precautionary principles and principles of prudent avoidance into action, and for the general public to investigate ways to minimise risks at home and at work.

Reducing exposure

-8-

How to reduce exposure to EMR at home

Where not so long ago our forebears cooked in a wood-fired stove, laundered clothes in a boiler and hand-turned wringer, and sewed by candlelight, their modern counterparts are surrounded by an amazing array of electronic wizardry. There is the electric carving knife to dissect the roast, the electric toothbrush to massage the gums, the mixmaster, the dishwasher, the microwave, and the personal computer. Without doubt, this electronic gadgetry has improved the speed and ease with which we attend to our chores, and has increased our comfort and productivity. But it has changed the nature of our living and working environment to the extent that we are now enmeshed in a complicated tapestry of current-carrying threads.

These icons of an affluent society all emit electromagnetic radiation, sometimes to the detriment of our health and well-being. In 1988 US researcher Dr. Elizabeth Hatch investigated the link between using household appliances and the incidence of childhood leukemia. She found that risk of the disease was increased by mothers who used electric blankets during pregnancy, and with the frequency with which children used hairdryers, video arcade machines, and television video games. It also increased with the frequency of television viewing, regardless of the distance children sat from the television. (Hatch 1998.) For many people, appliances are not used for sufficiently long periods to cause a real problem. For others, who are allergic to electricity, using an electric iron can cause them to feel unwell.

Reducing exposure is merely a sensible precaution for some people; for others, it is a vital necessity.

It is essential to reduce the exposure to EMR of children and foetuses in particular. Cells are most vulnerable to radiation when they are growing and dividing; because this occurs more often with young children and foetuses, they are especially vulnerable. It is extremely important that children do not sleep near a source of high fields and do not spend long periods of time near or playing with appliances that emit EMR.

Because electromagnetic fields drop off rapidly with distance, the further you are from the source, the less exposure you are receiving. You can avoid spending time in dangerous 'hot spots' by judicious arrangement of furniture. Don't place beds or favourite chairs next to sources of high electromagnetic radiation such as meter boxes, computers or an electric storage hot-water system (even if these are on the other side of the wall) or against walls in which there are cables, water pipes, dimmer lights or switches for ceiling fans, because EMR travels through walls.

Particularly high fields are generated by appliances that contain transformers such as digital clocks. These include videos, electric ranges, microwave ovens, and some alarm clocks. Appliances that contain electromagnets (such as phones, televisions, and fridges) can also be a problem, particularly for people who are allergic to electricity. Finally, appliances with motors—such as washing machines or fridges—produce higher fields than those without.

As a number of serious complaints have been found at levels of just 3 mG, it may be wise to ensure that continuous exposure is kept below this level.

Many of the following measurements are given for the lowest, highest, and mean exposures, as measured by the US EPA (1992.)

Accessories

Metal accessories—jewellery, watches, glasses etc.—act rather like an antenna, both absorbing and reflecting radiation. To reduce this effect when you are working in exposed situations for lengthy periods (for example, using a computer at work), it may be wise to wear glasses with plastic rather than metal frames. *See also 'Watches'.*

Aeroplanes

A US survey of magnetic fields on an aircraft flight deck (Nicholas et al.) showed substantial and fluctuating fields. (Fields on aircraft are generally above 300 Hz, rather than the 60 Hz of the power system.) Mean exposure over a number of flights was approximately 17 mG. Although emissions ranged from 2.3 mG to 160.4 mG, most (90 per cent) of the measurements were between 6 and 30 mG.

As well as considering the effects of EMR from the plane on people, there is also the question of the effects of radiation-emitting equipment carried by passengers, on the aircraft's delicate electronic equipment. As discussed in chapter four, all Australian airlines disallow the use of mobile phone during a flight, and the Civil Aviation Safety Authority is considering regulations to ban their use from all flights in the future.

Alarms

Many alarm systems use infra-red sensors which detect body heat. Those that are fed by a double-insulated mains transformer and are not earthed can give off high fields, but this can be reduced by earthing. However, infra-red itself is not suspected, and usually the devices are situated away from householders.

Anti-theft systems

These shoulder-height pedestals, looking rather like gates, are installed at the entrance to many supermarkets and variety stores to deter would-be thieves by issuing an alarm when unscanned merchandise passes between them.

The pedestals transmit a time-varying magnetic field, commonly around 8.2 MHz. If unscanned merchandise passes between them, the emissions cause the metal or glass compound in the tag or the strip underneath the barcode to vibrate at the same frequency as the pedestal. This vibration is detected by the pedestals which respond by issuing an alarm.

The systems do not just interact with inanimate merchandise from the store. Several studies have found that anti-theft systems interact with pacemakers and other implanted medical devices, and there are reports of people passing out when standing close to them. (McIvor 1998; Santucci 1998; Sridha 1998; Zipes 1999.)

In the US, the Food and Drug Administration recommended in September 1998 that doctors advise patients with implanted medical devices to take the following precautions:

- not to remain near the anti-theft systems longer than necessary and not to lean on them.
- to be conscious of the fact that anti-theft devices may be hidden; and
- to ask security personnel not to hold hand-scanners near implanted devices;

Because most people pass quickly through the gates, they are usually not a problem. However, it would be wise not to stand too close to them if you are working in a store or while waiting at the checkout.

Baby-crying monitors

Because these devices emit electromagnetic fields, it is wise to locate them at least one metre away from a child's bed.

Fields of 4-15 mG have been measured at 15 centimetres from a monitor and fields of 0-2 mG at 30 centimetres. (US EPA.)

Barcode scanners

These devices, found in most supermarkets and libraries and many warehouses, convert encoded data to intelligible information. In supermarket scanners a narrow laser beam passes over the barcode, and its light is reflected to a sensor. The sensor detects how long the beam took to pass over each segment of the barcode, and this information is translated electronically, and conveyed to the computer. Hand-held library scanners use visible or infrared light rather than laser.

Dr. Gene Sobel conducted some measurements of these devices. He found that the hand-held 'guns' did not produce any EMR, whereas those embedded in supermarket benches produced fields of between 3 and 35 mG. It is likely that the fields emanate from the motors, and that the position of the motors accounts for the difference in fields in different models.

The devices are more likely to be a concern for workers than the public.

Batteries

Battery-operated equipment causes less problems than equipment powered by mains electricity because it produces a direct—rather than alternating—current. However, some people are allergic even to these fields.

Bedrooms

Because you spend many hours there, it is extremely important that you are not exposed to high levels of electromagnetic radiation in the bedroom. Make sure the bedroom does not contain any unnecessary electrical appliances (such as televisions), and keep appliances as far as possible from the bed.

Beds

Because metal-framed beds and metal-spring mattresses are good conductors, it is important to remove all possible sources of EMR from near the bed. Don't place the bed against a wall where there are cables, water pipes, dimmer switches, a meter box, or ceiling-fan control box. Make sure there are no electrical cables, powerboards or, especially, copper pipes running under the bed as these may conduct electricity and thus produce EMR. And try not to locate other metal objects (which conduct electricity) near the bed.

Some people believe that the springs of a bed can behave like an antenna, receiving and re-radiating signals that affect the sleeper. You can discover whether a mattress's springs have become magnetised by EMR quite simply. Place a compass on a chair beside the bed and reverse the mattress end to end. If the position of the compass needle changes, the mattress has become magnetised. You can demagnetise a bed using a degausser (such as those used to demagnetise tape decks). Plug the degausser into the power system and move it slowly away from the bed. However, the degausser produces high fields, and the process needs to be repeated regularly (perhaps every six months).

Water beds can also radiate EMR and should be avoided if possible.

If you are in the market for a new bed, you could consider buying one with a wooden frame and a futon or feather mattress.

See also 'Electric blankets'.

Cars

Unfortunately, not even the open road provides escape from the electromagnetic buzz of everyday life, as cars in motion can emit quite considerable electromagnetic fields. Swedish researchers Kjetil Vedholm and Dr. Yngve Hamnerius measured fields of up to 50 mG in the rear of a car traveling at 60 mph. (Vedholm and Hamerius 1997.) A study by Milham et al. found fields of 20 mG in the back seat of a car, and these did not change even when the car was coasting with its engine turned off. (Milham 1999.)

Fields are sometimes produced from steel-belted radial tires in motion which generate an alternating magnetic field which differs even among virtually identical tires. The field can be reduced by degaussing the tires every few months using a hand-held magnetic tape degausser. Samuel Milham used a Geneva Audio/Video Tape Eraser, Model PF211 which he held close to the hand-spun tires of a car that had been jacked up. After degaussing a tire in this way, the field it produced dropped from 20 to 2 mG, and stayed low for several months. (Letter to *MW News*, March/April 1998.)

Late-model cars also contain an array of electronic wizardry powered by a DC system (which, remember, may be less biologically active than the AC fields of the power distribution system). The fields from these devices can be reduced by not operating equipment such as windscreen wipers, radios, and heating unnecessarily. A US survey found that electric and petrol-driven cars have similar field levels—which emanate mostly from the tires and auxiliary systems—and that it is possible to design electric cars with lower magnetic fields than conventional cars. (Whittemore 1998.)

Mobile-phone use in cars

It is extremely important not to use a mobile phone inside a car unless the car has an external aerial because the metal body of the vehicle reflects the signal backwards and forwards, exposing passengers to the radiation.

Quite apart from the radiation they emit, mobile phones have been found to have other health impacts when used in a car. A number of studies have shown that use of a mobile phone while driving, whether or not it is hands-free, increases the risk of an accident (see chapter four).

Ceiling fans

Fields of up to 50 mG have been measured at 30 centimetres from a ceiling fan, with a median measurement of about 3 mG. Fields of 6 mG have been measured at 60 centimetres. (US EPA.)

If the control to the ceiling fan is located near the bed, make sure that the fan is turned off before going to sleep, or else move the bed.

Clocks

Digital alarm clocks contain a transformer, and hence generate relatively high electromagnetic fields. An alternative is a battery operated bed clock, which has a negligible field, or a wind-up clock.

Clothing

Clothing from natural fibres such as cotton, wool, linen, hemp, or silk is slightly conductive, and allows static electricity to leak to earth. Synthetic fibres are much less conductive, allowing them to 'store' static electricity which can be uncomfortable for the wearer.

As the wiring in bras can absorb and reflect radiation, it may be better to avoid underwire styles. The British medical paper *Doctor* reported that women who wear bras with underwire support tend to have a greater incidence of breast lumps. (*Electromagnetic Hazard and Therapy* 1997.)
See also 'Accessories'.

Computers

Most of the radiation from a computer is from the screen or Visual Display Unit (VDU) which produces Extremely Low Frequency fields, radiofrequency radiation and electrostatic fields, x-rays, ultraviolet radiation, infrared radiation, and ultrasound emissions, although some of these occur in small quantities. (US Department of Health and Human Services 1999: 55.) Older computers had higher emissions and were less well shielded than their more recent counterparts, most of which now comply with the more precautionary Swedish standard MPR2 guidelines.

Frequent computer use has been associated with eye problems, skin problems (redness and blotchiness), headaches, memory loss, nasal problems including sinusitis, tiredness, dizziness, numbness in arms and legs, facial pain (blisters, metallic taste in mouth), breathing problems and heart palpitations, joint pains, and allergies to electricity. (SIF 1996.)

Swedish researcher Dr. Olle Johansson has found cellular and neural changes in people who have developed skin sensitivity from VDU use. These include a high number of immune-reactive cells, abnormal patterns of nerve fibres, a larger number of mast cells, and changes in distribution of neuropeptides in the skin. Measurable changes occurred in the skin after subjects were exposed to electromagnetic fields, indicating that the effects of exposure are real and not to be dismissed as psychosomatic. (Johansson et al 1996.)

Several studies have linked VDU use to an increase in premature and still births, birth deformities, and miscarriages, as you have seen in chapter five. Moreover, the blue light from the VDU may also cause damage to the retina and lens of the eye (Ham 1983), and increase the risk of cataracts. (Zaret 1976) Since these studies, many computer manufacturers have voluntarily reduced emissions by complying with the recommended Swedish guidelines.

If your computer flickers, it may be in a high magnetic field. According to John Lincoln's experience, computers begin to flicker at about 15 mG.

Laptop computers are worth a special mention. Because a laptop's screen uses a liquid crystal display, it does not emit electric, and magnetic fields in the same way as conventional screens. Relatively high fields can sometimes be measured at the keyboards, however, because they are situated above the computer's power unit. If this is a problem, you could attach a separate keyboard. The battery charger, which contains a large transformer, will also be generating rather high fields, so try to keep it away from you as you work.

To reduce your exposure:
- Remember that most of the emissions emanate from the sides and rear of a computer, and locate the equipment where it will have the least impact (not on the other side of the wall from a bed, for example). Moving the keyboard so that the operator is at least one metre from the VDU reduces exposure significantly.
- Place the hard drive as far away from you as possible. If you have a tower type of hard drive, make sure it is not positioned near your legs.
- Don't rest a laptop computer on your legs while working for long periods.

- Purchase a field-reducing screen which sits over the front of the VDU. Make sure that it has an earth connection that is attached to the metal frame of the computer.
- Try to limit the time your family spends using the VDU, and discourage children from playing computer games.
- Make sure both the VDU and the operator are earthed. In the latter case, wear natural fibres and leather-soled shoes (or bare feet).
- Because metal can re-radiate EMR, you may prefer not to wear metal accessories (see *accessories*).
- If buying a computer, choose a brand that boasts lower emissions. Most computers now comply with the more precautionary Swedish standard (MPR2 Guidelines), that produce a magnetic field of 2.5 mG or less at about half a metre from the screen.
- Unplug the computer when it is not in use.

Cordless phones

Conventional cordless phones radiate only a fraction of the power of mobile phones, as they need only transmit to a nearby antenna situated within the home or office. Peak power levels range from 5 to 25 milliwatts (mW) and is often less than 1 mW, compared to peak power levels of 500 mW to 1 watt for mobile phones. For this reason, cordless phones have been generally thought to be safe, although no research has been conducted into the potential health effects of using a cordless phone.

The latest model cordless phones operate at a much higher power, similar to that used by many mobile phones. Phones using Spread Spectrum technology, often called CDMA, operate at a frequency of 902-928 MHz, close to the mobile-phone band, and have a peak power of up to 400 mW. These phones differ from their lower-power predecessors by virtue of their ability to transmit messages longer distances and to reconstitute the signal.

Recently, British scientist Dr. Gerard Hyland claimed that cordless phones may have a worse impact on health than mobile phones because they are used more frequently and for longer periods of time. 'You could say these are worse than mobiles because you have the phone and the base station, both emitting microwave radiation, sitting in the same room with you.' (*Adelaide Advertiser,* 6 July 1999.)

One study that has relevance to the question of the safety of cordless phones is a case study by Lennart Hardell of a woman who developed a rare cancer of the scalp (angiosarcoma). The woman had used a cordless mobile phone for more than an hour a day for ten years on the left side of her head, and later a GSM mobile phone for a few minutes a week. The tumour developed in the part of the head that had been most highly exposed to microwaves from the phones. (Hardell 1999.)

Defibrillators
See 'Pacemakers'.

Dental

Amalgam fillings can be a problem for some people. These fillings are comprised of roughly 50 per cent mercury (a neurotoxin, enzyme inhibitor, and source of free radicals), 25 per cent tin, 25 per cent silver, with traces of copper, zinc, and paladium. The combination of different metals and conductive saliva forms a simple battery which produces a tiny current. Exposure to EMR—and perhaps the amalgam battery itself—can trigger the release of mercury from the amalgam, and this is the case especially with computer users. Removing amalgam fillings has proven beneficial for some people, but has triggered allergies in others. (A study by Örtendah and Högstedt (1995) found that 'EMF-exposure more than doubled the mercury release from amalgam' in divers working with under-water welding equipment.)

Like amalgam fillings, braces can generate a small current in the body which has sometimes been found to adversely affect behaviour.

Digital clock radios

Emissions from these are often around 8 mG. (Mine measures 100 mG.) Keep them at least one metre from the bed. If you have a clock radio built into the bed-head, *stop using it and unplug it.*

Dimmer switches
See 'Lights'.

Dryers

To avoid fields from the electric clothes dryer, use the dryer as little as possible and make sure you don't over-dry the clothes.

Fields of 2–10 mG have been measured at 15 centimetres from dryers and fields of 0–3 mG at 60 centimetres. (US EPA.)

Electric blankets

Electric blankets are particularly dangerous because of their proximity to the body and the fact that they are often left on all night. Because looped cables give off higher fields than straight cables, an electric blanket, when turned on, can produce a very high field.

Measurements of over 200 mG have been found within 15 centimetres of older-style electric blankets. Newer styles have reduced emissions which generate fields of less than 20 mG at the same distance. However, I believe that *any* electromagnetic radiation should be avoided at night, and recommend that you use the electric blanket to warm the bed only, then turn it off, and *unplug the cord from the wall.*

A study by Savitz (1990) found that children of pregnant women who had used electric blankets had 2.5 times the risk of developing brain tumours. Dr. Kathleen Belanger later found that women who used electric blankets at the time of conception had a slightly increased risk of spontaneous abortion. (Belanger 1998.) Finally, a study by G.M. Leel found that the use of electric blankets affected the levels of melatonin, which, as we have seen, plays a role in protecting the body from cancers. (Leel 1999.)

Electric shavers

The high fields emitted by these shavers—4 to 600 mG at 15 centimetres and 0 to 100 mG at 30 centimetres—are dangerously close to the head. (US EPA.)

It could be prudent to use a re-chargeable battery-operated shaver, a safety razor (or grow a beard!). However, keep in mind that electric shavers are used for only relatively short periods of time.

Electronic pest controls

These operate by emitting an electromagnetic signal that is claimed to be intolerable to a pest. This has the potential to interact with the human body—particularly small organs and glands—and increases the level of electrosmog in the home environment.

I have received reports of people suddenly experiencing adverse health effects after such a device was installed in their homes. Given that, in our experience, some units actually attract the pests, their value may not be worth the risk to health.

Furnishings

People with intolerances to static fields often select furnishings from natural fibres such as cotton, wool, linen, hemp or silk. Some prefer to avoid synthetics which facilitate static electricity or treat them with anti-stat spray.

Hairdryers

The high fields emitted by a hairdryer are precariously close to the head, even though they may only be experienced for short periods at a time. Even so, it may be wise not to use a hairdryer unnecessarily and to hold it away from your body if using a dryer professionally.

Fields of up to 700 mG have been measured at 15 centimetres from hairdryers and fields of up to 70 mG at 30 centimetres. (US EPA.)

Hands-free kits (see Mobile phones)

Several studies have reached different conclusions on the merits of these devices.

In March 2000 the British Consumer Association, *Which*, tested two models of hands-free kits. It found that the kits acted as aerials, conducting three times the usual level of radiation directly into the head. The report concluded, 'Think again if you use a hands-free kit to protect yourself from mobile-phone radiation—the two we tested increase the radiation levels inside your head compared with holding the phone by your ear.'

Other studies that have measured specific absorption rate or SAR (the amount of radiation absorbed into a gel-filled mould representing the head) have found a protective effect. In August 2000 *Choice* Magazine found that hands-free kits reduced SARs by 92 per cent at the head, but that SARs at the waist (next to the phone) were high. These results are similar to those of a study commissioned by *New Scientist* in April 1999 which found that a hands-free kit reduced exposures by 94 per cent.

If you decide to use a hands-free kit, it may be wise to take the following sensible precautions: first, hold the mobile phone away from the body to avoid irradiating internal organs; second, limit your use of the mobile phone as much as possible.

Heating

Though all electric heaters produce electromagnetic fields, the risk varies depending on distance from the heater and length of exposure. To reduce exposure, keep the heater at a distance and use only when necessary.

Bar heaters
These produce a relatively small field, depending on the type, and are not really a problem at a distance. We have found that a typical field may be about 10 mG at the heater on a low setting, reducing to about 0.5 mG around a metre away, which is about the background level of domestic EMR. The field is greater at higher settings.

Ceiling heaters
The location of the motor in the ceiling provides ample separation from residents below, so that they are not being exposed to high fields.

Underfloor heaters
Elements are arranged in a grid pattern below the floor, so that fields are distributed throughout the system only centimetres from the body. In a brand new, very expensive home with underfloor heating throughout, John Lincoln measured a field of 170 mG at floor level. At waist level, where a tiny baby sat cradled in its mother's arms, he measured a field of 90 mG. In another home, he measured an incredibly high 639 mG at the floor, and found that this was generating a field of 5 mG in a child's bed in the floor below. In our opinion, these levels are dangerously high.

I would advise against spending much time in areas heated in this way.

Underfloor hot-water heating systems
These systems heat water that is contained in a grid of pipes below the floor. Though much safer than underfloor elements, they have the potential to conduct return currents through the pipes. If using such a system, have the pipes assessed for conductivity.

Hot-water systems
Some systems generate quite high electromagnetic fields. If you use an off-peak model, the heater will be operating during the night while you are in bed, so it is important that your bed is situated away from both the hot-water heater and the meter box, and any water pipes from the heater.

Induction cooktops
These cook food by producing an alternating magnetic field under the saucepan to induce heat. The fields they emit are roughly the same as for a conventional

cooktop. Induction cooktops operate at higher frequencies (22,000 to 34,000 Hz) than the power system.

Irons

An iron can generate a field of 6–20 mG at 15 centimetres, which is about the distance that the hand is from the appliance during use. (US EPA.)

You can reduce your exposure by using a cordless iron that heats when connected to a stand but is not connected with the power system while it is being held for ironing.

Kitchen appliances

Although all of these appliances generate an electromagnetic field, they are often not used for long periods of time. It may be prudent, however, to avoid unnecessary appliances and to position yourself away from appliances that are operating for long periods.

The following fields have been measured from common kitchen appliances by the US EPA:

blenders	30–100 mG at 15 centimetres
	5–20 mG at 30 centimetres
can openers	500–1500 mG at 15 centimetres
	40–300 mG at 30 centimetres
coffee makers	4–10 mG at 15 centimetres
crock pots	3–9 mG at 15 centimetres
	0–1 mG at 30 centimetres
dishwashers	10–100 mG at 15 centimetres
	6–30 mG at 30 centimetres
food processors	20–130 mG at 15 centimetres
	5–20 mG at 30 centimetres
garbage disposal units	60–100 mG at 15 centimetres
	8–20 mG at 30 centimetres
mixers	30–600 mG at 15 centimetres
	5–100 mG at 30 centimetres
electric ovens	4–20 mG at 15 centimetres
	1–5 mG at 30 centimetres
electric ranges	20–200 mG at 15 centimetres
	0–30 mG at 30 centimetres

refrigerators   0–40 mG at 15 centimetres

                                      0–20 mG at 30 centimetres

toasters   5–20 mG at 15 centimetres

                                      0–7 mG at 30 centimetres

See also 'Microwave Ovens'.

Lights

The form of lighting which produces the lowest electromagnetic fields is the humble incandescent filament lamp.

Fluorescent lights

A fluorescent light results in a high EM field in the range of 20 to 100 mG (US EPA) which can affect people living or working directly above or below it.

Fluorescent lights are also dangerous because

- they contain toxic polychlorinated biphenyls (PCBs);
- they emit harmful, carcinogenic electromagnetic radiation like that from VDUs; and
- they emit ultraviolet radiation.

London's Dr. Valerie Beral found that people who worked under fluorescent lights experienced a higher rate of malignant melanomas. (Beral 1982.) It is possible to reduce the number of fluorescent lights used and to improve the quality of light from them by using Silverlux Reflectors (see 'Resources').

Halogen lamps

These increasingly popular lights emit ultraviolet radiation and contain transformers which emit high electromagnetic fields. John Lincoln has measured 700 mG from a halogen desk lamp that was plugged in and turned on at the socket, but not at the lamp. When the lamp itself was turned on, the reading reduced to 500 mG. (When the lamp was turned off only the primary coil was operating. Turning the lamp on brought the secondary coil into circuit, which cancelled some of the field from the primary coil.)

Quartz halogen lights

These emit more ultraviolet light than do glass globes.

Low-energy lights

These lights emit radiation in a range of frequencies.

Low-voltage lights

These contain transformers which emit high electromagnetic fields and also emit ultraviolet radiation.

Light dimmers

Dimmers contain transformers which emit high electromagnetic fields.

Meter boxes

John Lincoln has found that a meter box typically produces fields of 100 mG at the box and 30 mG on the inside of a brick cavity wall. For this reason, it is not a good idea to place a bed or a favourite chair on the internal wall next to the box. Try not to sleep against a wall which contains a meter box, even if box and bed are on different storeys of the building.

Microwave ovens

Microwave ovens operate at a frequency of 2,450 MHz, in the microwave range of the electromagnetic spectrum. They heat food by energising the oxygen component of water molecules.

A number of studies have found that cooking food in a microwave oven adversely affects the food and the consumer. Richard Quan (1992) found that heating breast milk in a microwave oven reduces its immunologic properties. Hans Hertel and Bernard Blanc (1994) found that the blood of people who had eaten food cooked in a microwave oven showed significant changes, consistent with deteriorating health, including a decrease in haemoglobin values, a decrease in (good) cholesterol, short-term decrease in while blood cells and an increase in leukocytes. Japanese researcher Dr. Fumio Watanabe (1998) found that cooking food in a microwave oven reduces the amount of B12 (which protects against neurological problems and heart disease) more quickly than does conventional cooking.

It is important to realise that microwave ovens are legally allowed to leak radiation of up to five milliwatts/cm^2 at a distance of five centimetres or more from the oven. As they age, the ovens can leak additional radiation, often as a result of deteriorating door seals. (*NHMRC* 1985.) If you use a microwave oven, make sure that door seals are in good condition, and check the oven for leakage.

As well as microwave radiation, microwave ovens give off high electromagnetic fields at 50 Hz. John Lincoln has measured fields of 65 milligauss from ovens plugged in but not in use, and fields of 250 milligauss from ovens during cooking time.

Dangerous bacteria can survive in food cooked in a microwave. Bacteria are more likely to survive when food is reheated in a microwave than in a conventional oven. (Ashton and Laura 1997.)

Furthermore, the packaging designed for microwaving foods—waxed bags and plastic film—has been shown to contaminate food during cooking. (Ashton and Laura 1997.) Recently a teenage high school student in the US, Claire Nelson, won the American Chemical Society's top science prize for a project to investigate whether or not toxins in the plastic wrap migrated into the food during cooking. She covered virgin olive oil with four types of plastic wrap, microwaved them and investigated the contents of the oil. Nelson found that suspected carcinogen di(ethylhexyl)adepate (or DEHA) seeped into the food, contaminating it at 200 to 500 parts per million (compared to the US standard of 0.05 parts per billion). She also found that xenoestrogen migrated into the food—significant, because this chemical is responsible for low sperm counts and breast cancer. (Associated Press 3 April 2000.)

To reduce your exposure to the radiation from microwave ovens, you can:

*   use a conventional oven where possible;
*   regularly check microwave leakage (appropriate devices are available from electronics stores);
*   avoid standing next to a microwave oven during cooking, and don't install it at head height or underneath a bench that is frequently used;
*   turn off the oven when it is not in use, and remove the plug from the socket; and
*   allow food that has been cooked in a microwave oven to stand for a few minutes before eating it, to allow cells to normalise (as recommended by manufacturers).

Mobile phones

A significant amount of the radiation emitted by a mobile phone is deposited in the user's head. Not surprisingly, then, using mobile phones has been associated with a wide range of symptoms including headaches, tingling sensations in the head, heat or pressure in the temple, ear ache, bleeding ears, eye problems including distortion of vision, memory loss, fatigue, high blood pressure, brain tumours, DNA damage, changes to size, shape, and growth of cells and changes to DNA, lymphomas in rats, changes in the brain's electrical activity, and learning problems.

While manufacturers continue to proclaim the safety of their products, it is interesting to note that at least five companies have taken out patents for low radiation phones. Meanwhile, insurance giant Lloyds of London has refused to insure manufacturers against health risks to customers.

"Don't worry Mike. All it means is that mice shouldn't use mobile phones."

In order to reduce your exposure to radiation from mobile phones, you can:

- use conventional phones in preference to mobile phones;
- reduce the amount of time spent on a mobile phone;
- make sure the antenna is extended during calls;
- hold a mobile phone about 3–4 centimetres from the head during a call and don't hold the phone next to the head while dialling and waiting for an answer;
- install an external aerial on the car, but *ensure that it is not located near the rear window of a car, particularly near a baby capsule or child seat.*
- avoid using a mobile phone inside a train, bus, or car (without external aerial) as the metal frame of the vehicle reflects signals back into the body of the user and fellow passengers;
- avoid carrying a mobile phone next to the body, as the phones emit radiation even when not in use;
- keep mobile phones away from children as children absorb more radiation than adults, and do ensure that youngsters don't suck the aerial; and
- choose lower-power CDMA phones.

As explained earlier, the emissions from a mobile phone are measured according to how much radiation a gel-filled plastic model of a human head absorbs. This is known as the Specific Absorption Rate or SAR. The Australian, the United Kingdom and the United States governments are shortly to make SAR measurements available to the public inside packaging (Australia) or support literature (US).

However, SAR measurements are a guide, but not necessarily an assurance of safety. For example, mobile phones with low SARs have been found, by many

users, to cause more symptoms than mobile phones with higher SARs.
See also chapter four.

Mobile-phone shields

In response to community concern about the safety of mobile phones, a range of shielding devices is now being marketed. These include pouches, earpieces, protective amulets, and devices claiming to radiate a 'protective' field. There is presently no uniform and independent testing of these devices, and consumers are obliged to rely on manufacturers' pronouncements of efficacy. Until such time as independent testing is available, prospective buyers are encouraged to be aware of the risks.

Even if a shield does reduce the radiation from a mobile phone, there may still be problems for the user from the low frequency fields of the handset battery. One study found an increased rate of death among chicken embryos exposed to the radiation from a mobile phone with a device that reduced the microwave emissions. (Youbicier-Simo 1999.) This may be because a shielding device obliges the phone to work harder to establish contact with the base station, and results in higher surges of current from the battery.
See 'Handsfree kits'.

Night lights

Night lights emit small electromagnetic fields which John Lincoln has found to be around 3 mG, so it is important that these not be located next to a child's bed.

Ovens

Electric ranges can generate a field of 20–200 mG at 30 centimetres. (US EPA.)

To reduce your exposure, it may be wise not to stand close to the operational stove while you prepare food. Often high fields are generated by the digital clock display in the range, and this can be disconnected if the field is a concern. Gas ovens emit lower fields, and may be a worthwhile choice if you are particularly allergic to EMR and are in the market for a new range.

Pacemakers

A review of scientific studies by Dr. Antonia Sastre found that EMR from power sources can interfere with implanted pacemakers and defibrillators at levels of just

2 G or 1.5 kV/m. (Sastre 1998.) Numerous studies have found that some mobile phones—both analogue and digital—also cause interference (see chapter four).

Personal Communication Systems (PCS)

This offshoot of mobile telephony offers the versatility of cordless phones that function as a mobile phone beyond the home. These are really mobile phones operating at higher frequencies of 1800 MHz which require lower-power antennas to be erected at short distances throughout the community, sometimes on lampposts. This means that people are exposed to these emissions wherever they might be. Health effects have already been reported in areas where the system has been introduced.

Pets

Whereas cats prefer to spend time in areas where there is higher EMR, dogs do not. Watching where your pet sleeps may give you an indication of spots to avoid.

Power points

Whether or not an appliance is plugged in, a powerpoint will generate an electrical field. We have found these fields to vary from 10 to 300 volts/per metre. If building, you can reduce the electrical field from a powerpoint by encasing the power cable in earthed steel tube for a metre or so on each side of it. However, unless you spend long periods of time directly against the powerpoint, this is probably not of great concern, as the fields drop off rapidly the further away you are.

Railways

In New South Wales, railways use DC (Direct Current) power, which is less biologically active than the AC (Alternating Current) of the power system. However, when a DC field intersects at right angles to an AC field, there is opportunity for a resonance effect (described in chapter six) which can affect ions in the body. Theoretically, this could occur when DC railways and AC powerlines intersect.

The fields from a railway are similar to those from an ordinary domestic powerline. However, they also put out a radiofreqency signal.

See also 'Trains'.

Refrigerators

The refrigerator produces a field of up to 40 mG at 15 centimetres and up to 20 mG at 30 centimetres, and the highest fields are usually located at the bottom and back of the unit. (US EPA.) Avoid working next to it or sitting/sleeping directly on the other side of the wall from it, particularly given that—unlike most other appliances—the electric motor is running intermittently.

Remote controls

A remote control sends a very low power and very brief radio signal to a television set, car, or garage door to initiate action.

Using a remote control for the television or stereo means that the appliance is always turned on and that electric fields remain high. After using the device, ensure that the appliance itself is turned off, preferably at the wall, and unplug it if possible.

Satellite dishes

Because the power from the satellite is low, these dishes need to be large enough to collect sufficient signal. Domestic dishes are receivers only, and so do not generate any radiofrequency fields. They should not, therefore, pose a health risk to neighbours.

Sewing machines

A number of studies suggest that the radiation from sewing machines constitutes some risk to health.

Professor Claire Infante-Rivard reported that the use of sewing machines by pregnant women, either at home or in industry, can increase the chance of their children developing leukemia (1995). Dr. Eugene Sobel of California found an increased incidence of Alzheimer's disease among people exposed to high electromagnetic fields at work, including seamstresses (1995).

Home sewing machines have been found to produce fields of 12 mG at chest level and 5 mG at head level. (On once occasion, John Lincoln measured 90 mG at chets level.) Industrial sewing machines have been found to produce fields of up to 35 mG at chest level and 215 mG at knee level. (EPA 1992: 40.)

Static electricity

Static electricity, produced by friction or movement, is present in equipment such as hair dryers, dishwashers, washing machines, air conditioners, computers, and photocopiers. It can also occur when people walk across a nylon carpet, use a nylon brush or wear synthetic clothing. While, for most people, static discharges are largely inconsequential, others experience severe stress and/or pain.

For those people who experience such reactions, it is advisable to select natural fibres for clothing, footwear, and household furnishings. These items can also be sprayed with anti-stat sprays. Some shoes are now being marketed with anti-stat pads in the heels. Walking barefoot on the ground earths the body, and can relieve feeling of discomfort from static buildup.

Stereos

These produce emissions of around 3–5 mG. Don't sit too close to them while they are operating, and unplug them from the power point when they are not being used.

Telephones

While it doesn't emit an electromagnetic field as does an operating appliance, the telephone can nevertheless affect the user. The earpiece of a telephone contains electromagnets, and the mouthpiece contains a microphone which can cause discomfort for those with severe allergies to EMR.

Sometimes a telephone line can pick up a field from power cables, particularly if the two are laid close together. This can result in extremely high electric fields in the line and at the earpiece, which can cause the user extreme discomfort. This can be addressed initially by installing a speaker phone so that the user is at a comfortable distance from the phone and does not need to hold the receiver against the head. Ultimately, it may be necessary to re-route the phone line, ensuring that it does not come into contact with any wiring, or else use shielded phone cable.

The FEB of Sweden (an association of people affected with allergies to electricity) suggests a number of cheap innovations to reduce discomfort from using a telephone. You can cut a small hole in the base of a plastic cup and attach this to the receiver to keep the phone away from the head. Alternatively, attach the cone of a stethoscope to the loudspeaker inside the phone's receiver to place even greater distance between the speaker and the phone. Speaker phones which also distance the user from the receiver are available commercially.

Televisions

According to the US EPA, these produce emissions of around 3–12.5 mG, though John Lincoln has found fields of 30 to 130 mG at the sides and backs of sets. Because, like VDUs, they contain a cathode ray tube, they also emit Extremely Low Frequency fields, radiofrequency radiation, electrostatic fields, x-rays, and a small amount of gamma-radiation. TVs emit higher fields than VDUs, particularly newer models which contain a switch-mode power supply instead of a conventional transformer, and this creates an additional high frequency (50–150 kHz) signal.

For this reason, don't sit too close to them while they are operating (US author, Lucinda Grant, 1992, recommends a distance of three metres), and unplug them from the power point when they are not being used. Because a television can produce higher fields at the rear than in front of the screen, it is important that it not be placed on the other side of the wall from a bed.

Because VDUs emit lower fields than televisions, some people choose to watch television via their computer. This requires a multimedia computer with a UHF receiver, a television card, and an antenna.

The transmission of high-definition digital television (HDTV), which began in January 2001, may pose an additional risk to the broader community. During the eight-year transition to digital television both systems will be operating simultaneously, resulting in an increase in transmission power in the community. When analogue transmission is discontinued, these levels will initially drop, because digital transmission requires less power than analogue, but then they will rise as more channels are introduced.

Thyristors

Normally, alternating current can only be used for fixed speed devices. Use of a thyristor—a type of electronic switch—allows the device to operate at variable speeds, as discussed in chapter two. Thyristors switch the signal on/off in the middle of a wave, producing peaks and transmitting the disturbed signal into the wider distribution network. Appliances which contain them tend to produce higher than normal fields.

Thyristors are contained in variable-speed drills, electric mixers, hair dryers, light dimmers, ceiling- and kitchen-fan controls, and in an increasing number of more modern appliances.

Trains

Electric trains give off quite high levels of EMR, and this can affect both passengers and drivers. Measurements in the US found an average field of approximately 125 mG at seat height and peaks up to 500 mG. (*National Research Council* 1996). I have measured fields of up to 50 mG on trains in Sydney.

Several studies have found that railway workers who are exposed to high levels of electromagnetic radiation experience increased rates of cancer (Floderus 1994), double the average risk of lymphocytic leukemia (Alfredsson 1996), and have a significantly high number of chromosome breaks. (Nordenson 1996.) (*MW News* July/August 1996.)

The use of mobile phones on trains has been a contentious issue. Because the metal carriage reflects the signal, the phone irradiates not just the user but also other passengers. They can also be extremely annoying, as any peak hour traveller would know. Mobile phones have been banned in some trains in Tokyo—because of the nuisance impact of many conversations competing loudly with background noise. (*Electronic Telegraph* 25 March 1997.)
See also 'Railways'.

Transformers

Like its larger counterparts in the power grid, a domestic transformer changes the voltage of the electricity as appropriate for a particular appliance. They are found in a variety of appliances. As well as halogen lights, they are found in digital clock displays in clock radios, microwave ovens, and kitchen ranges. In some cases the transformer is obvious as a box-like attachment to a battery charger, radio, etc.

Transformers generally produce quite high levels of EMR: John Lincoln has measured a field of 1,000 mG at the case of a typical battery recharger.

Many flats and offices are situated above the large transformers that are often located in the basemenst of large buildings. Because these can generate extremely high fields, it is important that they be well-shielded. If you live or work near a transformer of this sort, it might be wise to arrange for an EMR survey to measure the fields.

Vacuum cleaners

Like all appliances, vacuum cleaners generate electromagnetic fields. Fields of 100–700 mG have been measured at 15 centimetres from the motor; fields of 20 to 200 mG at 30 centimetres; and fields of 4–50 mG at 60 centimetres. (US EPA.)

VDUs

See 'Computers'.

Videos

Videos emit significant fields while on, so make sure the recorder is turned off properly (not just with the remote control switch, which leaves the device still functional), and preferably unplug it at the wall when it is not being used.

Walkie talkie

Walkie talkies emit radiation with greater power than mobile phones. A study on the effects of radiation from walkie talkies on hospital equipment found that walkie talkies caused interference at a distance of 10 metres, and accordingly advised care in hospital and emergency situations. (Irnich and Tobisch 1999.)

Presumably, it would be wise to take similar precautions in connection to the delicate machinery of the brain. Because the antenna is close to the user's head, it would be prudent to use a walkie talkie only when necessary, and not for extended periods. By implication then, they should not be used as toys by children.

Washing machines

If the washing machine is used at night, try to ensure that it is not placed on the other side of the wall from an occupied bed.

Of course I'd love to do the housework, darling, but there's my health to think of.

Fields of 4–100 mG have been measured at 15 centimetres from washing machines and fields of 1–30 mG at 30 centimetres. The highest field seems to be produced during the spin cycle. (US EPA.)

Watches

Battery-operated watches can interfere with the body's natural electromagnetic fields and are a source of discomfort to people with allergies to EMR. If buying a watch, you may like to avoid this problem by choosing a wind-up design.

Water pipes

In many homes electricity chooses a course through water pipes to complete its circuit to earth, and sometimes the fields in one home run to earth in a neighbour's water pipes. If this is the case, avoid sleeping near conductive pipes or insert a segment of plastic pipe to eliminate the fields (see chapter ten).

Nancy Wertheimer found that children living in homes where water pipes conducted elevated fields had a higher than average risk of cancer (1995).

Workshop

All electric tools generate electromagnetic fields which vary enormously and drop off rapidly with distance. In some situations you will be able to reduce your exposure by working away from an appliance that is operating, such as a battery charger.

The following fields have been measured from:

*battery chargers	3–50 mG at 15 centimetres
	2–4 mG at 30 centimetres
drills	100–200 mG at 15 centimetres
	20–40 mG at 30 centimetres
power saws	50–1,000 mG at 15 centimetres
	9–300 mG at 30 centimetres

(US EPA.)

(*As mentioned previously, John Lincoln has measured typical fields of 1,000 mG at the case of a battery charger.)

Fields from soldering irons of between 26 and 63 mG were measured by the State Electricity Commission of Victoria in 1992. (*Report of the Panel on Electromagnetic Fields and Health to the Victorian Government* 1992, p. 50.)

What else you can do

- Because children and foetuses are particularly vulnerable to EMR, make sure they do not play with appliances unnecessarily, and keep them a safe distance from televisions, etc.
- If you live near powerlines, you can screen your home from some of the electrical field by planting trees between the house and the powerlines. However, trees—and other objects—do not screen out magnetic fields.
- Avoid buying gimmicky appliances you don't need (such as electric toothbrushes).
- Buy low-emissions appliances. Creating a demand will improve the availability of such appliances.
- Turn off appliances that are not presently in use and pull the plug from the socket if you are near them for long periods (appliances that are turned off but not unplugged continue to generate a small electric field).
- Install a demand switch which cuts off power to a circuit when that circuit is not being used. This could be used to remove power from cables next to a bed at night.
- As an alternative to the demand switch, run a new power circuit to the bedroom with its own master switch. This, too, would allow mains power to be cut off to the bedroom at night without affecting appliances elsewhere in the house.
- To release the buildup of static electricity from your body, earth yourself by walking barefoot over the bare ground.
- Ensure that your diet includes antioxidants, which play a role in reducing radiation damage to cells. Common antioxidants are vitamins C and E, the mineral selenium, the herb chaparral, and the enzyme superoxide dismutase. However, it would be wise to consult a naturopath before making dietary changes.
- Shield sources of high EMR exposure (see chapter ten).

Reducing your family's exposure to EMR doesn't have to be a complex engineering feat of daunting magnitude. It can be as simple as moving the digital alarm clock from the bedside table or unplugging the electric blanket before retiring for the night. Many of the suggestions outlined in this chapter require little or no expense and are easy to implement. Yet they can be enormously potent. Reducing EMR exposure may not only help avert some of the more serious illnesses described in earlier chapters, but may improve your family's sense of energy and well-being. It won't hurt to try.

-9-

How to reduce exposure to EMR at work

A s our reliance on time-saving equipment grows, so does our exposure to electromagnetic radiation in most occupations. Not only have computers become an essential part of virtually all businesses, but there is a profusion of other sources of EMR. Supermarkets and variety stores have introduced anti-theft devices emitting radiofrequency EMR; automotive workshops have heat sealing equipment which emits radiofrequency EMR; libraries have degaussers that emit magnetic fields. Drivers are exposed to EMR from vehicles; hairdressers to EMR from dryers; electricians to fields from powerlines; welders to radiofrequency EMR from the equipment they use ...

No wonder so many of us feel tired by the end of the working day.

Reducing exposure at work has advantages for all concerned. Workers benefit from improved health and a sense of well-being. Governments benefit by a reduction in medical benefit claims. Employees benefit by improved productivity and—as has happened in Sweden, where allergies to EMR have resulted in loss of many to the work force—by retaining a valuable bank of expertise. Moreover, employer's efforts to protect their workers reduces their liability for compensation claims from EMR-injured workers should EMR be proven in the future—as seems likely—to cause health problems.

Does EMR at work affect employee's health?

From the strong association between EMR and health problems that has been described in preceding chapters, it would be surprising if people working in high fields for long periods of time were not experiencing some symptoms. Indeed, some studies have found that people exposed to high levels of EMR at work have a greater risk of disease than workers in less exposed occupations, as you have seen in the previous chapters. People working in electrical utilities have been

157

found to have higher rates of brain cancer, leukemia, and breast cancer. Workers exposed to high fields from a variety of sources have found increased risks of breast cancer, heart problems, reproductive problems, and Alzheimer's disease.

These findings are consistent with surveys that have been conducted in a number of workplaces. In 1992 Professor Bengt Knave conducted a survey of 731 employees at five large Swedish companies. He found that more than one employee in seven suffered from allergies to EMR. The symptoms most commonly experienced were skin and nervous problems. The former included the experience of red or pink patches, rashes, blushing, prickly sensations, aches, tightness, itching, and sensitivity to light. Nervous symptoms included dizziness, prickly sensations, fatigue, weakness, headaches, breathing problems, perspiration, depression, heart palpitations, and forgetfulness. Many employees reported these symptoms in connection with computer use, but others had experienced problems from using machinery or proximity to fluorescent lights or wiring. On the positive side, the study found that workers with such allergies who reduced their exposure were able to continue working. (*Forskning and Praktik* 1992.)

In 1993 and 1996 the Swedish Union of Clerical and Technical Employees in Industry (SIF) conducted surveys of members' health. (SIF 1996.) It reported the following symptoms associated with the use of electricity in the office, the incidence of which had doubled in the intervening three-year period:

- eye problems—difficulty in seeing, smarting pain, feeling of grit in eyes;
- skin problems—dryness, redness, blotchy complexion;
- nose problems—stuffed nose, runny nose and sinusitis;
- stinging face—face feels hot, swollen, stings and blisters appear;
- mouth problems—sores, blisters and metallic taste;
- facial pain—pain over entire face, pain concentrated in teeth and jaws;
- mucous membrane problems—dry mucous membranes and abnormal thirst;
- headaches, memory loss, and depression;
- abnormal tiredness and difficulty concentrating;
- dizziness, faintness, and nausea;
- breathing problems and palpitations;
- joint pains, especially in shoulders and arm; and
- numbness in arms and legs, cramp, and pricking sensation

Certainly some of these symptoms are being experienced by Australian workers, judging by calls I've received from workers in high-exposure situations. Unfortunately, some callers report that employers have not considered that EMR exposures may play a role and have tended to dismiss the effects as 'stress'. As greater awareness about the health effects of EMR develops, it will be possible to diagnose the problems early and to take appropriate action to protect workers.

What do you mean my memory and concentration aren't up to scratch? You're the one who gave me the mobile phone last Christmas!

To what are workers being exposed?

The most common sources of EMR in the office are—in order of effect—fluorescent lights, photocopiers, office wiring, and computer monitors. (Florig and Barry 1992.) Emissions from these can produce a range of effects that have been found to affect health and productivity.

However, high electromagnetic fields are not limited to the office. They are found wherever electrical equipment is present and in most occupations. In the United States, the National Institute for Occupational Safety and Health (NIOSH) compiled information about the exposures that workers received in a variety of occupations, which is summarised in the table below. The list provides the average magnetic field exposure received by a worker in a day.

EMF Measurements Averaged Over a Workday

Industry and Occupation	ELF magnetic fields measured in mG	
	Median for occupation	Range for 90% of workers*
Employed men in Sweden		
Construction machine operators	0.4	0.2– 0.6
Motor vehicle drivers	0.8	0.3– 1.9
Teachers in theoretical subjects	1.2	0.4– 3.1
Machine repair and assembly personnel	1.7	0.3– 3.7
Retail sales staff	2.7	0.8– 4.4
Electrical workers in various industries		
Electrical engineers	1.7	0.5–12.0
Construction electricians	3.1	1.6–12.0

TV repairers	4.3	0.6– 8.6
Welders	8.2	1.7–96.0
Electrical utilities		
Clerical workers without computers	0.5	0.5– 1.6
Clerical workers with computers	1.2	0.3– 6.3
Line workers	2.5	0.5–35.0
Electricians	5.4	0.8–34.0
Distribution substation operators	7.2	1.1–34.0
Workers off the job (home, travel, etc.)	0.9	0.3– 3.7
Telecommunications		
Installation, maintenance, and repair technicians	1.6	0.9– 3.1
Central office technicians	2.1	0.5– 8.2
Cable splicers	3.2	0.7–15.0
Auto transmission manufacturing		
Assemblers	0.7	0.2– 4.9
Machinists	1.9	0.6– 28.0
Hospitals		
Nurses	1.1	0.5– 2.1
X-ray technicians	1.5	1.0– 2.2
Garment industry workers in Finland		
Sewing machine operators	22.0	10.0–40.0
Other factory workers	3.0	1.0– 6.0

*This range is between the 5th and 95th percentiles of daily average measurements for an occupation.

Information courtesy National Institute of Environmental Health Sciences, National Institutes of Health.

Particular locations in a work place may have higher fields than others. Transformers, power cables, degaussers, and high-current equipment, for example, can expose workers to fields of hundreds of milligauss, so are best avoided where possible. The following table lists spot measurements that were conducted by NIOSH in a variety of workplaces. Most of the measurements were taken at the worker's location, though some were taken at fixed distances from the equipment, as indicated in the third column. In some cases the investigator took a series of measurements at different locations in the room, and these are referred to in the chart as 'walk-through' surveys.

EMF Spot Measurements

Industry and Sources	ELF magnetic fields measured in mG	Comments	Other frequencies
Mechanical equipment used in manufacturing			
Electric resistance heater	6000–14,000	Tool exposures	VLF
Induction heater	10–460	measured at	High VLF
Hand-held grinder	3000	operator's chest	
Grinder	110		
Lathe, drill press, etc.	1–4		
Electrogalvanizing			
Rectification room	2000–4600	Rectified DC	High static fields
Outdoor electric line and substation	100–1700	current (with an ELF ripple) galvanizes metal parts	
Aluminum refining			
Aluminum pot rooms	3.4–30	Highly-rectified DC current (with an	Very high static field
Rectification room	300–3300	ELF ripple) refines aluminum	High static field
Steel foundry			
Ladle refinery ladle electrodes active	170–1300	Highest ELF field was at the chair of	High ULF from the ladle's
Electrodes inactive	0.6–3.7	control room	big magnetic
Electrogalvanizing unit	2–1100	operator	stirrer High VLF
Television broadcasting			
Video cameras (studio and min-cam)	7.2–24	Measured 1 ft. away	VLF
Video tape degaussers	160–3300		
Light control centres	10–300	Walk-through	
Studio and newsrooms	2–5	survey	
Telecommunications			
Relay switching racks	1.5–32	Measured 2 in.-3 in. from relays	Static fields and ULF-ELF transients
Switching rooms (relay & electronic switches)	0.1–1300	Walk-through survey	Static fields and ULF-ELF transients
Underground phone vault	3–5	Walk-through survey	

Hospitals

Intensive care unit	0.1–220	Measured at	VLF
Post-anesthesia care unit	0.1–24	nurse's chest	VLF
Magnetic resonance imaging (MRI)	0.5–280	Measured at technician's work locations	Very high static field, VLF and RF

Government offices

Desk work locations	0.1–7	Peaks due to laser
Desks near power center	18–50	printers
Power cables in floor	15–170	
Computer center	0.4–6.6	
Can opener	3000	Appliance fields
Desktop cooling fan	1000	measured
Other office appliances	10–200	6 in. away
Building power supplies	25–1800	

Information courtesy National Institute of Environmental Health Sciences, National Institutes of Health.

Reducing exposure at work

The first step in reducing your exposures at work is to identify the source of the field. Often underfloor cables, transformers in an adjoining room or fluorescent lights to the floor beneath will not be obvious at first glance, so there are merits in obtaining an EMR survey of the workplace. Having identified the source of the fields, you may be able to shield them or to simply move work stations further away.

Some of the common sources of EMR at work—and suggestions for reducing exposure from them—are listed below.

Anti-theft systems
See p. 133

People employed to check bags of customers leaving a shop would be advised to stand at a distance from the pedestals.

Because the gadget that removes the tag from clothes contains a strong magnetic field, it should not be held close to the body.

Barcode scanners
See p. 134

To reduce the exposure to workers operating the scanners, the devices should be measured to ascertain the location of the highest fields, and these areas should be shielded or avoided.

Computers
See p. 136

Remember that most emissions emanate from the sides and rear of a computer, so try to locate the equipment where it will have the least impact. Avoid banks or clusters of computers where workers are exposed to radiation from their own and other machines. One US computer-shielding company recommends a distance of 2.10 metres between computers. (Grant 1992.)

Computers can represent a risk not just to a company's employees, but also its customers. Some people have reported discomfort from standing next to the rear of a computer in a bank or post office. Take care when you arrange your office so that customers are not exposed in this way, and try to allow a separation of at least two metres between the customer and the computer.

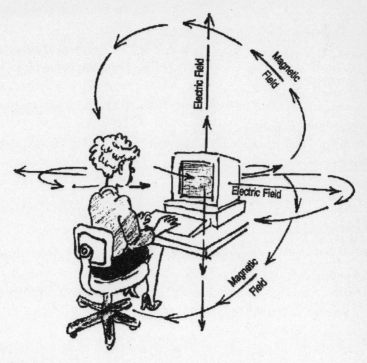

Howard Saunders, Labor Institute; reprinted with permission.

Fax machines

Fields of 4–9 mG have been measured 15 centimetres from fax machines. (US EPA.) To avoid exposure, do not situate a fax machine on or near an employee's desk.

Fluorescent lights
See p. 144

To reduce your exposure, replace fluorescent lights with ordinary incandescent bulbs or full spectrum lighting. Alternately, you can reduce the number of fluorescent lights used in your building while increasing the quality of light from them by using a Silverlux Reflector (see 'Resources'). Make sure that work stations are situated away from fluorescent lights, particularly those from the floor below. (The locations of these lights are easily detected by measuring the magnetic field at floor level.)

Replace fluorescent tubes if they begin to flicker, as there is evidence that the brain is particularly sensitive to this effect. (Sandstrom et al. 1997.)

If you do need to use fluorescent lights, it is perhaps best to choose full-spectrum daylight tubes.

Photocopiers

These generate high levels of EMR as well as chemical emissions, so should be kept away from workstations.

Fields of 4–200 mG have been measured at 15 centimetres from copiers; fields of 2–40 mG 30 centimetres away; and fields of 1–13 mG at 60 centimetres, though fields of several hundred mG have been measured by Australia's CSIRO.

Powerlines

Workplaces situated in the second storey of a building may have powerlines located directly outside, just a metre or so from workstations. If this is the situation, employees may be working in extremely high fields. (John Lincoln has measured fields of over 200 mG in such a situation, at the desk of a worker who was seriously ill!)

To reduce exposure of employees, move workstations away from the powerline and/or ask the electric utility to install bundled conductors.

Radar guns

There is some evidence that police officers who used radar guns developed a higher rate of testicular cancer and melanomas (see chapter five).

A report for the US National Institute for Occupational Safety and Health suggested a number of measures that could be implemented to reduce microwave exposure by officers using radar guns:

- 'hand-held devices should be equipped with a switch requiring active contact to emit radiation. Such a switch, referred to as a 'dead-man switch', must be held down for the device to emit radiation, even though the electrical power to the device is on;
- older hand-held devices that do not have a 'dead-man switch' should not be placed with the radiating antenna pointed toward the body, whether it is held in the hand or placed near the officer. A holster or other similar device should be used as a temporary holder for the radar when not in use;
- when using two-piece radar units, the antenna should be mounted so that the radar beam is not directed toward the vehicle occupants. The preferred mounting location would be outside the vehicle altogether, although this may not be practical with older units that cannot withstand adverse weather conditions. Other options, for example mounting on the dashboard of the vehicle, are acceptable if the antenna is at all times directed away from the operator or other vehicle occupants. Mounting the antenna on the inside of a side window is not recommended;
- radar antennas should be tested periodically, for example annually, or after exceptional mechanical trauma to the device, for radiation leakage or back scatter in a direction other than that intended by the antenna beam-pattern; and
- each operator should receive training in the proper use of traffic radar before operating the device. This training should include a discussion of the health risks of exposure to microwave radiation, and information on how to minimise operator exposure.'

(Lotz et al. 1995.)

Supermarket scanners

see Barcode scanners

Telephones

See p. 151

Transformers

Whether in an appliance or the powergrid, transformers generate relatively high fields.

Many office buildings contain their own transformers, often located in the basement, and these can radiate fields through the walls to adjoining offices. The study by Samuel Milham (1996) showed that office workers located above a transformer had a higher rate of cancer over time. This unfortunate experience has been repeated far too often in the community.

If your office is located above a transformer, you would be advised to have the fields measured. Often it is possible to shield the fields, as you will read in the following chapter.

Guidelines to protect workers

Already unions and employers are responding to the need to protect workers from exposure to EMR. Both Sydney's Royal North Shore Hospital and Britain's Public and Commercial Services Union have developed guidelines designed to protect workers from exposure from mobile phones (see chapter seven). Unions in Australia and Sweden have also introduced guidelines to protect workers from EMR—principally from VDU and power-source exposures.

"You'll love working here, McCauley. It's a real buzz!"

ACTU guidelines

In Australia the Australian Council of Trade Unions (ACTU) has developed 'Guidelines for Screen Based Work' (May 1998) which are based on a policy of prudent avoidance. The policy recognises 'that even if the risk is small it will increase with growing use of electronic means of communication and administration.'

Many of its recommendations pertain to computer VDUs, for which it

suggests the following maximum emissions, measured at 50 centimetres from the VDU and 30 centimetres from the front of the screen:
- an electric field of less than 10 V/m at 5Hz to 2 kHz;
- an electric field of less than 1 V/m at 2kHz to 400 Hz;
- a magnetic field of less than 2 mG at 5Hz to 2 kHz; and
- a magnetic field of less than 0.25 mG at 2 kHz to 400 kHz (at 50 centimetres around the VDU).

It also recommends that the VDU have a low electrostatic potential to 'prevent dust particles moving from the screen to the user due to differences in potential.'

Because laptop computers can produce peaks (and also high keyboard fields), the ACTU suggests that they not be operated on a worker's lap.

The policy recommends an average exposure of workers to less than 2 mG over an eight-hour day, and suggests that this can be achieved by the following measures:
- purchasing improved equipment or shielding equipment;
- ensuring that VDUs and other equipment are adequately spaced so that combined fields do not exceed the standard; and
- rerouting or shielding cables.

Where these measures are impracticable, other options may include:
- limiting time worked in areas where the exposure standard is exceeded;
- locating workers outside high-exposure areas. In these circumstances, warning placards should set out the practices that should be observed in relation to the particular area or equipment; and
- switching off equipment which is not in use.

The ACTU guidelines take into account risks to children and foetuses. They state, 'Because of possible associations of EMFs with birth defects and miscarriage, consideration should be given to providing workers who are contemplating pregnancy in their families with the option of moving to duties other than screen based work.

Children may be more sensitive to EMFs than adults, so more stringent measurements and precautions should be taken if there is child care in the workplace. School computer rooms, which may use old equipment in crowded conditions, should also receive rigorous attention.'

Swedish Trade Union Confederation

In Sweden the high number of workers which has developed allergies to EMR, particularly after prolonged computer use, has robbed industry of a wealth of competent experience and has put pressure on employees and unions to protect worker safety.

The Swedish Trade Union Confederation (LOA) has responded with a booklet called *Cancer and Magnetic Fields at the Workplace* (1993) which suggests the following precautions they would like to see implemented to protect workers from EMF. They suggest that:

- 'the principle of caution be applied as regards exposure to magnetic fields;
- all unnecessary exposure be avoided;
- new places of work be designed and equipped in such a way that the exposure to magnetic fields is minimised;
- manufacturers of electrical equipment aim to minimise the magnetic fields;
- manufacturers give details of the levels of magnetic fields in connection with the sale of such equipment;
- no employee be exposed to an average exposure exceeding 0.2 microTesla (2mG) per working day;
- temporary high exposures be minimised as far as possible;
- the employer maps out the existing levels of magnetic fields and, when necessary, draws up plans of action in accordance with the internal control regulations;
- the staff in question be informed and trained;
- practical measures to reduce exposure be taken without further delay, such as indication of areas with high exposure, reduction of magnetic fields, transfer of work sites, and changed work organisation;
- the distribution of responsibilities between the different authorities be clarified;
- Swedish laws and ordinances reflect a viewpoint corresponding to the principle of caution mentioned above;
- the stray currents be eliminated by the introduction of five wire systems;
- the National Board of Occupational Safety and Health, pending the draft for a hygienic limit, issues a regulation in accordance with the views above; and
- the research on electric and magnetic fields be continued and intensified.

(LOA 1993, *Electrosensitivity News* 3:3:8.)

Other ideas for guidelines to protect workers

- Undertake a survey of fields in the workplace.
- Endeavour to keep fields below 3 mG as much as possible.
- Make information on field reduction available to workers.
- Purchase only low-emissions equipment.
- Purchase radiation-reducing computer screens.
- Make a landline available to all workers.

- Do not oblige employees to use mobile phones. Supply them with pagers, phone cards, or answering machines in preference.

Working in a safe environment is not a luxury. It is the fundamental right of all workers. Today's workers are the first generation to use the plethora of sophisticated technological gadgetry that has become basic to many occupations: the first to use computers, the first to use mobile phones, the first to use the internet, and the first to be surrounded by a maze of powerlines and equipment. Given that so many technological developments have occurred recently, we have hardly begun to experience the effects of this workstyle on our health. So whether or not health effects from EMR have been conclusively 'proven', it makes powerfully good sense to reduce worker exposure as much as possible.

A good place to start is by management and unions developing their own policies on reducing workplace exposures and by providing information to workers on how to reduce their risks. For more ideas along these lines, see chapter seven.

-10-

Designing to reduce and avoid emissions

While high fields from our power system can generally be reduced, doing so can sometimes be both expensive and inconvenient. Because, in many cases, problems are only detected after illness has developed, it can sometimes seem rather like shutting the stable door after the horse has bolted.

A far better approach is to reduce fields at the design stage, be it a house, an office, a subdivision or a city. This places responsibility on all those who play a role in the planning at any level: on governments and councils, on architects, on electricians and builders.

Choosing a building site

- Avoid selecting a site above an underground water supply. Some health problems seem to be caused by locating buildings above subterranean water supplies.
- Avoid an area where there is a geological break, as the action of the Earth's plates being squeezed results in piezoelectricity, which can result in health problems for occupants at the site.
- Much of the earth's crust contains naturally occuring radioactive radon, and radon seems to be attracted to high levels of EMR, such as those emitted by powerlines. Inside a building it could be worth ensuring that fields are kept as low as possible and that rooms are well ventilated to reduce the accumulation of radon.
- It may be useful to have the site assessed for EMR before building. John Lincoln has measured fields in excess of 5 mG from vacant blocks of land. These emanated in one case from an underground power cable, and in another from water pipes of a demolished house that were still connected to neighbouring pipes.

Construction materials

Natural building materials such as brick and stone are more porous than their modern counterparts of concrete, plastic, and glass. The former allow the building to 'breathe' admitting negative ions from outside.

Building design

- Locate transformers away from living or sleeping areas and workstations.
- If you line walls of the house with aluminium foil, ensure that the foil is properly earthed to divert the electric field component to earth.
- If building near a source of high EMR, locate bedrooms farthest from the source.

Wiring

- Use bundled conductors rather than two separate wires to connect the house wiring to the street supply.
- Locate the meter box away from high-use areas, especially the walls of a bedroom, where people spend many hours. Appropriate locations are the walls of a garage, cupboard, or wardrobe. Alternatively, mount the meter box on a separate galvanised steel pole away from the house.
- If the meter box must be on a bedroom wall, insert a sheet of steel (three millimetres thick) or mumetal between the box and the wall. The metal plate should be 20 centimetres longer than the height and width of the meter box, and must wrap around the sides of the box. Ensure that it is connected to earth.
- The ESAA suggests strategies for reducing radiation from the wiring associated with the meter box. It recommends:
  - 'Locating the main connecting wiring away from high-use areas in cases where meter location and switchboard location are separated by a significant distance ,for example where meters are installed at the fence and the switchboard is located at (or in) the house. The connecting wiring should be run with phases and neutral grouped together, and in a ceiling space rather than a wall space, for example.
  - Using service wires of insulated twisted construction, as they produce significantly less fields than open wire (bare conductor) construction.
  - Minimising or avoiding situations where heavy-current wiring, especially that of stoves and air-conditioning, is placed in wall cavities within the house. This type of wiring is best located and grouped together in the

ceiling [or the basement]. Close proximity of the phase wires and neutral helps to cancel the magnetic fields.
- In the case of two-way switches, running the neutral wire along the same path as the twin active wire connecting the two switches to provide a canceling effect on the magnetic fields.'

(Dolan et al.1998:19–20.)

- The location of the wires is enormously important, as the active wire usually generates an electric field of around 190 V/m.
  - *Do* locate electrical cables well away from areas where you will spend long periods of time e.g. where beds and chairs will be placed.
  - *Do not* locate wiring directly under the floor and avoid, where possible, running lengths of unshielded wire in the walls.
  - *Do* locate the wiring above the ceiling where possible, rather than under the floor, so that it is at the maximum distance from people.
  - *Do* locate wiring—whether it is in the floor or ceiling—above or below dividing walls, so that fields are not generated in living or working areas. This is particularly important when wiring is run close to sleeping areas.
- Make sure that active and neutral conductors are bound together, so that the field from one cancels the field from the other.
- To avoid the problem of high fields from the active wire, you can shield the mains cables by encasing the active and neutral in galvanised steel pipe or shielded metal conduit that is earthed.
- Alternatively, you can obtain shielded 240 volts mains cables from Belgium. (See 'Resources'.)
- Wiring is often earthed to water pipes, and this is obligatory in some situations. However, it can lead to high fields running through a house as we have seen. (See *Water Pipes*, below.)
- Wiring that is run next to a telephone cable can induce a field in the telephone line that can affect the user of the phone. If the phone line is used for a computer's internet connection, high fields can travel from the phone line into the computer, and hence to the keyboard. Ensure that wiring and phone line are run separately or shielded.
- It is also mandatory to connect the earth wire to an earth stake, in accordance with Australian Standard 3000. ('SAA Wiring Rules'.)
- Do not lay power cables in the same trench as copper water pipes as there is a risk that the pipes will conduct fields from the cables.
- When two switches are used to control one light, make sure that the three wires that connect the switches are run together, rather than separately.

Water pipes

Earthing the neutral wiring to the water pipes results in fields—often high fields—being conducted through the pipes, down walls, and underneath floors. If the pipes are located in an area where people spend time—particularly a bed, a workstation or perhaps a lounge room—people in the home can be exposed to fields well in excess of the 3 mG or 10 V/m 'safety net' and, as you have seen, this has led to many unfortunate experiences of illness.

Fortunately, however, this problem can be easily rectified. To reduce the flow of neutral current through the pipes, consult an electrician. He may advise you to engage a plumber to replace a section of the underground metal pipe with a small insert (10 centimetres or as desired) of plastic pipe. This is better done closer to the perimeter of the property than the house, so that the pipes continue to act as an earth, and thus provide better protection from electric shock. Also, ensure that the plumber packs the soil firmly around the join to avoid it subsequently rupturing.

High-rise buildings

While the view may be appealing, living or working in a high-rise building may have its disadvantages. Apartments or offices in line of sight of microwave transmitters may well have higher exposures to radiofrequency radiation. Moreover, the steel-frame construction appears to distort the earth's magnetic lines, conducting EMR and increasing levels within the building.

Most of the larger high-rise buildings contain their own transformers which convert higher-voltage power from underground cables to 240 volts for domestic or commercial use (see chapter three). Because transformers produce relatively high fields that radiate through walls and ceilings, people living or working in areas adjacent to or above the transformers can be at risk.

By shielding transformers and locating them away from work or living areas, much of this risk can be ameliorated. The ESAA has suggested a number of precautionary design measures for reducing fields from substations situated in high-rise buildings. These include:
- 'locating substations away from normally occupied areas such as offices, lunchrooms, etc.;
- planning the substation layout so that the low-voltage side is further away from adjacent offices, computer rooms, etc. than the high-voltage side;
- locating transformers, low-voltage busbars, disconnector switches, and other potentially large sources of magnetic field within the area of the substation as far away as possible from adjacent offices, etc.;

- if the floor above the substation is used as office space, avoiding where possible the direct ceiling mounting of heavy-current cables, open-type busbars or disconnector switches;
- locating all cable trays as far as possible from the substation ceiling and walls that separate it from adjacent offices, etc.;
- avoiding the laying of heavy-current cable directly on the floor of the substation if the floor below the substation is used for office space;
- designing busbars to minimise separation between phases and between phases and the neutral bus;
- if practicable, orienting transformers and other sources that have uneven field patterns so that their highest field-strength side is turned away from the field-sensitive area;
- where possible, using three-phase cables in preference to three single-phase cables;
- using a trefoil arrangement of cables when using three single-core cables in a three-phase configuration. In such cases, if the neutral conductor is a separate single-core cable, placing it, where practicable, in the centre of the trefoil formation of phases;
- selecting the substation equipment considering, among other important electrical parameters, its low magnetic-field design—that is 11,000/415 V distribution transformers in steel housings, compact metal-clad busbars;
- avoiding phase-by-phase grouping of single-core cables in parallel circuits; and
- distributing all large single-phase loads and all constant current load such as lighting and office equipment equally between three phases of the low-voltage supply.'

(Dolan et al.1998:17–18.)

Offices

In addition to transformers, many offices contain large switchboards which can also generate high fields. These, too, can be reduced by applying appropriate design methods. The ESAA recommends a number of precautions that can help achieve such a reduction. It suggests:
- 'Keeping the incoming line and associated meterpanel and/or busbars away from heavy-use areas. This will also help avoid computer interference problems;
- Avoiding the use of separate conductor trays for the energised and neutral wires. If separate trays are necessary, it is best to place them adjacent to

low/no use areas;
- Locating switchboards away from high-use office areas if possible;
- Locating workstations away from switchboards when laying out new or reorganised office areas. A distance of four to five metres is suggested to provide the additional benefit of avoiding computer VDU interference; and
- Using energy-efficient lift motors, air-conditioning equipment, and industrial motors and manufacturing equipment.'

(Dolan et al. 1998:17–19.)

One indicator of high fields in an office is a flickering computer screen (but there other reasons why this might occur as well). John Lincoln has noticed that screens begin to flicker at exposures of about 15 mG and are badly distorted at about 45mG. In many cases, he has been commissioned by companies to reduce fields simply to eliminate this interference with computers, without any thought of the likely interference such fields are having with their operators.

In one office, a transformer in the room below was generating high fields which were a problem to the employees and their equipment. To remedy the situation, John laid sheets of steel over the office floor under the carpet to block the fields. In another office, high fields were emanating from underground cables just outside the building. To shield these fields, John laid three layers of steel sheeting folded to create a tunnel over the offending cables.

Another source of computer interference—and therefore high fields—is the building lift, which can distort nearby VDU screens as it passes by. The culprit can be the lift counterweight. This weighty body is often constructed with redundant building material, including steel. As it travels between floors, the steel eventually becomes magnetised.

The solution is to demagnetise the counterweight which can be a difficult task.

Interestingly, many people report improvements to their health after the fields in their workplace have been reduced.

Shielding

As we have seen, 50 Hz fields electric fields are relatively easy to shield, and can be blocked by metals such as copper or aluminium. 50 Hz magnetic fields are more difficult to shield, and it is best to use a metal that contains iron.

Microwave radiation is much more difficult to shield. Because they are reflected from different surfaces, microwaves reach us from many directions. Like sunlight, they are quite pervasive and, in the same way as pulling down a blind on one wall doesn't totally screen out light from a room, so shielding one wall may not entirely keep microwaves at bay.

Fine-wire mesh provides some protection against UHF and microwave radiation. If the holes are smaller than the wavelength that is being received, the mesh reflects the signal. An example of its shielding properties is the fine mesh seen in the door of a microwave oven.

To shield against the radiation from a mobile-phone tower, you would need a wire mesh with a hole size of less than the size of the wavelength which, for a cellphone, is roughly a handspan. Thus most commonly available wire mesh, including chicken wire, can be used, and sometimes this is glued to a wall and subsequently covered with wallpaper. A metal flyscreen on the window can also help reflect signals.

Reducing exposures at a regional level

Given the evidence of health risks from EMR, it is incumbent upon all levels of government to apply precautions that will reduce exposure. Because many planning decisions are made at local level, there is an urgent need for local councils to develop strategies to protect the public.

Two councils which have contributed enormously to this process in Australia are Sydney's Sutherland Shire Council and Melbourne's Moreland Council.

Electricity pylons can make great carports. Some people have also found them convenient for drying the washing. However, we have to wonder whether previous administrators may have let us down by allowing high-voltage lines and houses to be built so close together.

Sutherland Shire Council, as previously mentioned, developed Australia's first policy on the siting of mobile-phone towers, a policy that was adopted by many other councils and which has enjoyed the confidence of the public and the co-operation of the carriers.

Moreland Council devised a thorough and well-researched strategy in 1998 on reducing public exposure, primarily to fields from power sources. It recommended that fields in its jurisdiction be reduced to 2 mG, and that this target be accomplished over a 20-year period. According to the document, 'The 2 mG limit has been chosen as it is the lowest practical field strength achievable—it is also the lowest level at which health effects have been associated with EMFs.' Many of its recommendations are included in the body of this book, and an excerpt from its strategy (as well as Sutherland Council's policy) is included in appendix B. Unfortunately, after several years, Moreland's strategy has still not been endorsed by the council. The document remains a promising—but impotent—ideal, a model for councils everywhere.

From recognising the merit of prudent avoidance and a desire to protect public health, a strategy for reducing public exposure is a natural evolution. Some suggestions for reducinge risks at a regional level are as follows:

- Encourage the local generation of electricity.
- Discourage reliance on the national power grid by supporting energy-efficiency initiatives and the greater use of solar power in cities.
- Ensure that buildings are not constructed under high-voltage powerlines by regulating for wide easements. Do not encourage children to congregate in easements by installing playground equipment or playing fields in them.
- Develop a policy which stipulates a maximum exposure level in the municipality of no more than 2 mG to be achieved over a period of time (see appendix B).
- Discuss with the local energy utility the possibility of upgrading and shielding existing substations.
- When electricity supply is being installed or upgraded, discuss with the energy utility methods for reducing fields. These could include:
  - locating transformers away from residences, classrooms, child-care centres, hospitals, and other sensitive areas;
  - using bundled conductors; and
  - undergrounding cables.
- Conduct an electromagnetic field survey of the area (the council area or area for proposed development, for example) so that high fields can be identified and plans developed to reduce exposure.
- Where possible, do not allow residences or child-care facilities to be con-

structed in areas where exposure is above 2 mG.

- Establish building guidelines that minimise fields using the recommendations suggested elsewhere in this chapter.
- Provide public information about the health risks of EMR and suggestions for reducing exposure at home and work.
- Develop a policy on the siting of mobile-phone base stations which considers the risks of EMR to health (see appendix B).
- Give careful consideration to approving construction of new buildings that are in line of site of existing mobile-phone base stations or other transmitters. Ensure that you obtain data on exposures that people would receive at all levels of the building.
- When approving applications for mobile-phone base stations, take into account the fact that metal structures—such as door frames, window frames, steel beams, and children's equipment will amplify the signals.
- Monitor research and other developments relevant to the health effects of EMR.

Electromagnetic radiation is no longer a fringe issue. It is a matter of vital and growing public concern. It touches every member of society except for those in the most remote locations. Thankfully its touch is most often impersonal—power for the vacuum cleaner or the printing press, thank you very much, and no further ado. However, increasingly that touch is intensely, acutely personal as people experience more and more symptoms—headaches and concentration problems from using mobile phones, sore eyes and fatigue from using a computer, or depression and immune dysfunction from power sources. And some, as we have seen, experience much worse: years of debilitating exhaustion, the suffering of cancer, the agony of a child with leukemia … all in people exposed to high fields.

Yet it doesn't have to be this way.

Governments can legislate to protect people, can introduce appropriate standards, can encourage innovative technology. Councils can introduce guidelines for new developments and building applications. Unions can devise policies to reduce worker exposure. Management can commission EMR surveys, can shield sources of EMR, can rearrange workstations to avoid hot spots, can buy radiation-reducing equipment, can reduce reliance on mobile phones—and can be rewarded with improved staff morale, productivity and health. Architects and builders can design and construct buildings to reduce unnecessary exposure. Homeowners can eliminate high fields in water pipes, can place beds judiciously, and can avoid high sources of exposure.

At every level of society we *can* make a difference.

And we *must* make a difference.

EMR is a now a public health issue of enormous proportions. While there is not yet conclusive scientific proof that it *causes* health problems, there is abundant evidence associating it with a host of difficulties, including the dreaded cancers and leukemias.

Proof is a luxury for which we cannot afford to wait. We do not need proof that a child will be run over before we introduce zebra crossings and 40 kmh speed zones outside a school gate. We do not need proof that a child will drown before we install a fence around our backyard pool. We did not need proof that tobacco was dangerous before we introduced warnings on cigarette packets. Must we have proof that EMR is dangerous before we take sensible precautions to reduce exposure?

Public health issues do not require us to wait for absolute, conclusive, and final proof before taking action. They require a sensible response to evidence; and, in this case, the evidence is considerable and mounting. Conclusive scientific proof will doubtless come, but it may be years away. In the meantime it is up to us all to take precautions to protect our family's, our workers', and our society's health.

It is our call and our privilege to be trustees of our children's future. As such, it behoves us to respond to the challenge set out in these pages; to do what we can to reduce their exposure now and to protect their health.

Appendix A

Effects of EMR on the body

Studies relevant to chapter four

Risk of using mobile phones while driving

Brown, 1969	Use of a mobile phone had a detrimental effect on controlling a car in 'non routine' situations.	Brown, I.D. et al., 'Interference Between Concurrent Tasks of Driving and Telephoning', *J. Appl. Psychol.* vol. 53, no. 5, pp. 419–24, Oct., 1969.
Brookhuis, 1991	Drivers' performance was affected by use of a mobile phone whether or not it was hands-free; users of hands-free 'showed better control over the vehicle' than those using a hand-held mobile phone.	Brookhuis, K.A. et al., 'The Effects of Mobile Telephoning on Driving Performance', *Accid. Anal. Prev.* vol. 23, no. 4, pp. 309–16, Aug. 1991.
McKnight, 1993	In simulated driving conditions, people distracted by mobile phone use were less likely to respond to driving situations. Non-response was greater in subjects over 50 years of age.	McKnight, A.J. and A.S., 'The Effect of Cellular Phone Use Upon Driver Attention', *Accid. Anal. Prev.* vol. 25, no. 3, pp. 259–65, Jun., 1993.
Alm, 1994	In simulated conditions, drivers using a mobile phone performed worse in easier driving conditions, with reaction time and speed adversely affected.	Alm, H. and Nilsson, L., 'Changes in Driver Behaviour as a Function of Handsfree Mobile Phones — a Simulator Study', *Accid. Anal. Prev.* vol. 26, no. 4, pp. 441–51, Aug., 1994.
Alm, 1995	In simulated conditions, drivers using a mobile phone performed worse in	Alm, H. and Nilsson, L., 'The Effects of a Mobile Telephone

	reaction time, particularly among older drivers. Use of the phone was shown to increase drivers' mental workload.	Task on Driver Behaviour in a Car Following Situation', *Accid. Anal. Prev.* vol. 27, no. 5, pp. 707–15, Oct., 1995.
Violanti, 1996	'Talking more than 50 minutes per month on cellular phones in a vehicle was associated with a 5.59-fold increased risk in a traffic accident.'	Violanti, J.M. and Marshall, J.R., 'Cellular Phones and Traffic Accidents: an Epidemiological Approach', *Accid. Anal. Prev.* vol. 28, no. 2, pp. 265–70, 1996.
Violanti, 1997	Drivers who used mobile phones were more prone to 'inattention, unsafe speed, driving on wrong side of road, striking a fixed object, overturning their vehicle, swerving prior to the accident, and running off the roadway.' Drivers using mobile phones had a greater risk of being killed in a traffic accident.	Violanti, J.M., 'Cellular Phones and Traffic Accidents', *Public Health* vol. 111, no. 6, pp. 423–8, 1997.
Redelmeier, 1997	Use of a mobile phone while driving quadrupled the risk of a collision, and a hands-free device offered no safety advantage.	Redelmeier, D.A. and Tibshirani, R.J., 'Association Between Cellular-telephone Calls and Motor Vehicle Collisions', *N. Engl. J. Med.* vol. 336, no. 7, pp. 453–8, Feb. 13, 1997.
Violanti, 1998	People who used a mobile phone while driving had approximately nine times the risk of a fatal accident, and the use of a mobile phone by anyone in the car doubled the risk of a fatal accident.	Violanti, J.M., 'Cellular Phones and Fatal Traffic Collisions', *Accid. Anal. Prev.* vol. 30, no. 4, pp. 519–24, 1998.
Lamble, 1999	Dialing numbers on a keypad reduced drivers' ability to detect a car ahead decelerating by about 0.5–1 second.	Lamble, D. et al., 'Cognitive Load and Detection Thresholds in Car Following Situations: Safety Implications for Using Mobile (Cellular) Telephones While Driving', *Accid. Anal. Prev.* vol. 31, no. 6, pp. 617–23, 1999.

Mobile phones interfere with pacemakers and cardiac defibrillators

Naegeli, B	Digital mobile phones interfere with pacemakers.	Naegeli, B. et al., 'Intermittent Pacemaker Dysfunction Caused by Digital Mobile Telephones', *J. Am. Coll. Cardiol.* vol. 27, no. 6, pp. 1471–7, 1996.
Hofgartner, 1996	28 models of pacemaker showed interference when exposed to mobile phones. 'Patients with implanted pacemakers should if possible not use mobile phones in the C [analogue] and D [digital] networks.'	Hofgartner, F. et al., 'Could C- and D-network Mobile Phones Endanger Patients with Pacemakers?' *Dtsch Med. Wochenschr* vol. 121 no. 20, pp. 646–52, 1996.
Barbaro, 1996	EMR from analogue phones interfered with a large number of pacemakers.	Barbaro, V. et al., 'Electromagnetic Interference of Analog Cellular Telephones with Pacemakers', *Pacing Clin., Electrophysiol.* vol. 19, no. 10, pp. 1410–18, 1996.
Irnich, 1996	Some mobile phone models caused interference with pacemakers.	Irnich, W. et al., 'Electromagnetic Interference of Pacemakers by Mobile Phones', *Pacing Clin. Electrophysiol.* vol. 19, no. 10, pp. 1431–46, 1996.
Chen, 1996	Some mobile-phone exposures—both digital and analogue—to patients with pacemakers caused interference.	Chen, W.H. et al., 'Interference of Cellular Phones with Implanted Permanent Pacemakers', *Clin. Cardiol.* vol. 19, no. 11, pp. 881–6, 1996.
Wilke, 1996	Two of 50 patients with pacemakers experienced interference from using mobiles: 'although interactions between cellular phone use and pacemaker function appear to be rare in our study, pacemaker dependent patients in particular should avoid the use of cellular phones.'	Wilke, A. et al., 'Influence of D-net (European GSM-Standard) Cellular Phones on Pacemaker Function in 50 Patients with Permanent Pacemakers', *Pacing Clin. Electrophysiol.* vol. 19, no. 10, pp. 1456–8, 1996.

Hayes, 1997	Mobile phones interfered with cardiac pacemakers when held next to them but not when held over the ear.	Hayes, D.L. et al., 'Interference with Cardiac Pacemakers by Cellular Telephones', *N. Engl. J. Med.* vol. 336, no. 21, pp. 1473–9, 1997.
Altamura, 1997	GSM phones interfered with pacemakers, and phones should not be carried close to the pacemaker.	Altamura, G. et al., 'Influence of Digital and Analogue Cellular Telephones on Implanted Pacemakers', *Eur. Heart J.* vol. 18, no. 10, pp. 1632–41, 1997.
Bassen, 1998	Digital mobile phones interfered with implantable cardiac defibrillators, with TDMA phones having greatest effect.	Bassen, H.I. et al., 'Cellular Phone Interference Testing of Implantable Cardiac Defibrillators in Vitro', *Pacing Clin. Electrophysiol.* vol. 21, no. 9, pp. 1709–15, 1998.
Schlegel, 1998	Some mobile phones interfered with pacemakers. 'The study ... supports the recommendation to maintain a separation distance of at least 6 inches [15 cm] between pacemakers and wireless phones.'	Schlegel, R.E. et al., 'Electromagnetic Compatibility Study of the In-vitro Interaction of Wireless Phones with Cardiac Pacemakers', *Biomed. Instrum. Technol.* vol. 32, no. 6, pp. 645–55, 1998.
Sakakibara, 1999	People with pacemakers reported interference from mobile phones.	Sakakibara, Y. and Mitsui, T., 'Concerns About Sources of Electromagnetic Interference in Patients with Pacemakers', *Jpn Heart J.* vol. 40, no. 6, pp. 737-43, 1999.
Trigano, 1999	Walkie-talkies and mobile phones interfered with some pacemakers.	Trigano, A.J. et al., 'Electromagnetic Interference of External Pacemakers by Walkie-talkies and Digital Cellular Phones: Experimental Study', *Pacing Clin. Electrophysiol.* vol. 22, no. 4 pt. 1, pp. 588–93, 1999.

Legend for subjects of research in following sections:

p = powerlines; mp = mobile phones; vdu = video display units; rf = radio frequencies; elf = extra low frequencies; vlf = very low frequencies; dc = direct current; rad = radar

Studies relevant to chapter five

Brain tumours

p	Lin, 1985	Electrical workers had a 2.8 higher rate of brain tumours.	Lin, R.S. et al., 'Occupational Exposure to Electromagnetic Fields and the Occurrence of Brain Tumors. An Analysis of Possible Associations,' *J. Occup. Med.*, vol. 27, no. 6, pp. 413–9, Jun., 1985.
p	Savitz, 1988	Children living in homes with high magnetic fields had a higher rate of brain cancer.	Savitz, D. et al., 'Case-control Study of Childhood Cancer and Exposure to 60-Hz Magnetic Fields', *Am. J. Epidemiol.*, vol. 128, no. 1, pp. 21–38, Jul., 1988.
p	Preston-Martin, 1989	People in exposed occupations had up to 4.3 times the usual rate of brain tumours.	Preston-Martin, S. et al., 'Risk Factors for Gliomas and Meningiomas in Males in Los Angeles County', *Cancer Res.*, vol. 49, no. 21, pp. 6137–43, Nov. 1, 1989.
p	Loomis, 1990	Electrical workers, telephone workers and electrical engineers had a higher rate of brain cancers.	Loomis, D.P. and Savitz, D.A., 'Mortality From Brain Cancer and Leukemia Among Electrical Workers', *Br. J. Ind. Med.* vol. 47, no. 9, pp. 633–8, Sep., 1990.
p	Juutilainen, 1990	Finnish workers exposed to ELF had a higher risk of brain tumours.	Juutilainen, J. et al., 'Incidence of Leukemia and Brain Tumours in Finnish Workers Exposed to ELF Magnetic Fields', *Int. Arch. Occup. Environ. Health*, vol. 62, no. 4, pp. 289–93, 1990.
p	Floderus, 1993	People exposed to prolonged high levels of occupational exposure had an increased risk of brain tumours above 2 mG.	Floderus, B. et al., 'Occupational Exposure to Electromagnetic Fields in Relation to Leukemia and Brain Tumors: a Case-control Study in Sweden', *Cancer Causes Control*, vol. 4, no. 5, pp. 465–76, Sep., 1993.

p	Savitz, 1995	Electric utility workers had an increased rate of brain cancer and risk increased with duration of employment and strength of magnetic field.	Savitz, D.A. and Loomis, D.P., 'Magnetic Field Exposure in Relation to Leukemia and Brain Cancer Mortality Among Electric Utility Workers', *Am. J. Epidemiol.*, vol. 141, no. 2, pp. 123–34, 15 Jan. 1995.
p	Guenel, 1996	Electric utility workers exposed to fields of 13 V/m for 25 years or more had a 7-fold increase in rate of brain tumours.	Guenel, P. et al., 'Exposure to 50-Hz Electric Field and Incidence of Leukemia, Brain Tumors, and Other Cancers Among French Electric Utility Workers', *Am. J. Epidemiol.*, vol. 144, no. 12, pp. 1107–21, 15 Dec. 1996.
p	Fear, 1996	Electrical and electronics industry workers had a higher rate of brain cancers and leukemias than controls.	Fear, N.T. et al., 'Cancer in Electrical Workers: An Analysis of Cancer Registrations in England', 1981-87, *Br. J. Cancer*, vol. 73, no. 7, pp. 935–39, Apr., 1996.
p	Rodvall, 1998	Occupations with high magnetic fields had an increased risk of brain tumours.	Rodvall, Y. et al., 'Occupational Exposure to Magnetic Fields and Brain Tumours in Central Sweden', *Eur. J. Epidemiol*, vol. 14, no. 6, pp. 563–9, Sep., 1998.
p rf	Thomas, 1987	Electronics manufacture & repair workers employed > 20 years had a 10-fold increase in astrocytic brain tumours.	Thomas, T.L. et al., 'Brain Tumor Mortality Risk Among Men with Electrical and Electronics Jobs: a Case-control Study', *J. Natl. Cancer Inst.*, vol. 79, no. 2, pp. 233–8, Aug., 1987.
p rf	Speers, 1988	Workers in transport, communications and electrical utilities industries had an increased rate of brain cancer.	Speers, M.A. et al., 'Occupational Exposures and Brain Cancer Mortality: A Preliminary Study of East Texas Residents', *Am. J. Ind. Med.*, vol. 13, no. 6, pp. 629–38, 1988.
p rf	Tornqvist, 1991	Assemblers/repairmen in radio and TV industry & welders had a	Tornqvist, S. et al., 'Incidence of Leukemia and Brain Tumours in

		2.9 times greater risk of brain tumours	some 'Electrical Occupations", *Br. J. Ind. Med.*, vol. 48, no. 9, pp. 597–603, Sep., 1991.
rf mp	Cleary, 1990	Glioma cells exposed for 2 hours to 27 or 2450 MHz EMR showed changes in proliferation.	Cleary, S.F. et al., 'Glioma Proliferation Modulated in Vitro by Isothermal Radiofrequency Radiation Exposure', *Radiat. Res.*, vol. 121, no. 1, pp. 38–45, Jan., 1990.
rf vdu	Ryan, 1992	Women who worked with cathode-ray tubes had an increased risk of brain tumour.	Ryan, P. et al., 'Risk Factors for Tumors of the Brain and Meninges: Results from the Adelaide Adult Brain Tumor Study', *Int. J. Cancer*, vol. 51, no. 1, pp. 20–7, Apr,. 1992.
rf	Grayson, 1996	Airforce personnel exposed to EMR had a small increase in brain tumours.	Grayson, J.K., 'Radiation Exposure, Socioeconomic Status, and Brain Tumor Risk in the US Air Force: A Nested Case-control Study', *Am. J. Epidemiol.*, vol. 143, no. 5, pp. 480–6, 1 Mar. 1996.
rf	Szmigielski, 1996	Polish military personnel had an increased rate of brain tumours.	Szmigielski, S., 'Cancer Morbidity in Subjects Occupationally Exposed to High Frequency (Radiofrequency and Microwave) Electromagnetic Radiation', *Sci. Total Environ.*, vol. 180, no. 1, pp. 9–17, Feb. 1996.
vdu	Beall, 1996	People who used VDTs for more than 10 years had an increase in brain tumours.	Beall, C. et al., 'Brain Tumors Among Electronics Industry Workers', *Epidemiology,* vol. 7, no. 2, pp. 125–30, Mar., 1996.
rf	Szmigielski, 1997	Polish military personnel exposed to RF had 2.7 times the average rate of brain cancer.	Second World Congress for Electricity and Magnetism in Biology and Medicine, Bologne, Italy, June, 1997. (*Microwave News* Jan./Feb. 98, p. 10.
rf mp	Hardell, 1999	A woman who used a cordless phone for 1 hour per day since	Hardell, L. et al., 'Angiosarcoma of the Scalp and Use of a

		1988 and a GSM mobile from 1994 on left ear developed a rare angiosarcoma at the point of highest EMR exposure from phone.	Cordless (Portable) Telephone', *Epidemiology* vol. 10, no. 6, pp. 785–6, 1999.
mp	Hardell, 1999	Though there was no significant risk of brain cancer, mobile phone users were overall 2.5 times more likely to develop tumours on the side of the head against which they held the phones.	Hardell, L. et al., 'Use of Cellular Telephones and the Risk for Brain Tumours: A Case-control Study', *Int. J. Oncol.*, vol. 15, no. 1, pp. 113–6, Jul., 1999.
rf mp	Hardell, 2001	People who used analogue mobile phones had a 26% increased risk of developing brain tumours and a 77% increased risk after 10 years.	Presented at Conference, 'Mobile Telephones and Health —the Latest Developments, London, June 6-7, 2001.

Leukemia

elf	Alfredsson, 1996	Swedish rail workers exposed to high EMR had double the average rate of lymphocytic leukemia.	Alfredsson, L. et al., 'Cancer Incidence Among Male Railway Engine-drivers and Conductors in Sweden, 1976-90', *Cancer Causes Control*, vol. 7, no. 3, pp. 377–81, May, 1996.
p	Wertheimer & Leeper, 1979	An increased risk of leukemia was found for children living near high electrical wiring configurations.	Wertheimer, N. and Leeper, E., 'Electrical Wiring Configurations and Childhood Cancer', *Am. J. Epidemiol*, vol. 109, no. 3, pp. 273–84, Mar., 1979.
p	Savitz, 1988	Children living in homes with high electric and magnetic fields from wiring configurations had an increased rate of leukemias and lymphomas.	Savitz, D.A. et al., 'Case-control Study of Childhood Cancer and Exposure to 60-Hz Magnetic Fields', *Am. J. Epidemiol.*, vol. 128, no. 1, pp. 21–38, Jul., 1998.
p	Coleman, 1989	People living within 25 metrres of a powerline had 1.3 times the risk of leukemia, and risk was greater for children under 18.	Coleman, M.P. et al., 'Leukemia and Residence Near Electricity Transmission Equipment: a Case-control Study', *Br. J. Cancer*, vol. 60. no. 5, pp. 793–8, Nov., 1989.

p	Juutilainen, 1990	Finnish workers exposed to EMR from power sources had an increased risk of leukemia.	Juutilainen, J. et al., 'Incidence of Leukemia and Brain Tumours in Finnish Workers Exposed to ELF Magnetic Fields', *Int. Arch. Occup. Environ. Health*, vol. 62, no. 4, pp. 289–93, 1990.
p	London, 1991	Children with leukemia were more likely to live near high electric wiring configurations and use appliances producing high EMR.	London, S.J. et al., 'Exposure to Residential Electric and Magnetic Fields and Risk of Childhood Leukemia', *Am. J. Epidemiol*, vol. 134, no. 9, pp. 923-37, 1 Nov. 1991.
p	Feychting & Ahlbom, 1993	Children living within 300 metres of 220 and 400 kV power lines had 2.7 times the risk of leukemia.	Feychting, M. and Ahlbom, A., 'Magnetic Fields and Cancer in Children Residing Near Swedish High-voltage Power Lines', *Am. J. Epidemiol.*, vol. 138, no. 7, pp. 467–81, 1 Oct. 1993.
p	Fajardo-Gutierrez, 1993	Children living near EMR from power sources had an above average risk of leukemia and the risk was more than 2.5 times the norm for living near high-voltage lines.	Fajardo-Gutierrez, A. et al., 'Residence Close to High-tension Electric Power Lines and its Association With Leukemia in Children', *Bol. Med. Hosp. Infant Mex.*, vol. 50, no. 1, pp. 32–8, Jan. 1993.
p	Floderus, 1993	Exposed workers in Sweden had a greater risk of leukemia and risk increased with exposure levels.	Floderus, B. et al., 'Occupational Exposure to Electromagnetic Fields in Relation to Leukemia and Brain Tumors: a Case-control Study in Sweden', *Cancer Causes Control*, vol. 4, no. 5, pp. 465–76, Sep., 1993.
p	Matanoski, 1993	Telephone linesmen with above average exposure had a risk of leukemia 2.5 times the average, and risk increased with exposure.	Matanoski, G.M. et al., 'Leukemia in Telephone Linemen', *Am. J. Epidemiol.*, vol. 137, no. 6, pp. 609–19, 15 Mar. 1993.
p	London, 1994	Electrical workers had a slightly increased risk of leukemia.	London, S.J. et al., 'Exposure to Magnetic Fields Among Electrical Workers in Relation to Leukemia Risk in Los Angeles

County', *Am. J. Ind. Med.*, vol. 26,
no. 1, pp. 47–60, Jul., 1994.

p	Theriault, 1994	Electric utility workers with higher than average exposure had an increased risk of leukemia.	Theriault, G. et al., 'Cancer Risks Associated With Occupational Exposure to Magnetic Fields Among Electric Utility Workers in Ontario and Quebec, Canada, and France: 1970–1989', *Am. J. Epidemiol.*, vol. 139, no. 6, pp. 550–72, 15 Mar. 1994.
p	Fear, 1996	Electrical workers had a higher rate of brain cancers and leukemias than controls.	Fear, N.T., et al., 'Cancer in Electrical Workers: An Analysis of Cancer Registrations in England, 1981–87', *Brit. J. Cancer*, vol. 73, no. 7, pp. 935–39, Apr., 1996.
p	Miller, 1996	Electric utility workers with high electric and magnetic field exposures had an increased risk of leukemia.	Miller, A.B. et al., 'Leukemia Following Occupational Exposure to 60-Hz Electric and Magnetic Fields Among Ontario Electric Utility Workers', *Am. J. Epidemiol.*, vol. 144, no. 2, pp. 150–60, 15 Jul. 1996.
p	Michaelis, 1997	Children living in homes with bedroom exposures of > 2 mG were twice as likely to develop leukemia. Children under four exposed to these levels had 7 times the average risk.	Michaelis, J. et al., 'Childhood Leukemia and Electromagnetic Fields: Results of a Population-based Case-control Study in *Cancer Causes Control*, vol. 8, no. 2, pp. 167–74, Mar., 1997.
p	Linet, 1997	Children exposed to EMR at home had no increase risk of leukemia at exposures of/below 2 mG, but increased risk at 3 mG or above.	Linet, M.S. et al., 'Residential Exposure to Magnetic Fields and Acute Lymphoblastic Leukemia in Children', *N. Engl. J. Med.* vol. 337, no. 1, pp. 1–7, 3 Jul. 1997.
p	Feychting, 1997	People exposed to EMR of over 2 mG at home had an increased risk of leukemias; those exposed at home & work had 3.7 times	Feychting, M. et al., 'Occupational and Residential Magnetic Field Exposure and Leukemia and Central Nervous

		the risk.	System Tumors', *Epidemiology*, vol. 8, no. 4, pp. 384–9, Jul., 1997.
p	Li, 1997	Children living within 100 m of high-voltage power lines had a risk of leukemia twice the average.	Li, C.Y. et al., 'Residential Exposure to 60-Hertz Magnetic Fields and Adult Cancers in Taiwan', *Epidemiology*, vol. 8, no. 1, pp. 25–30, Jan., 1997.
p	Li, 1998	Children living within 100 m of high-voltage powerlines had an elevated risk of leukemia.	Li, C.Y. et al., 'Risk of Leukemia in Children Living Near High-voltage Transmission Lines', J. *Occup. Environ. Med.*, vol. 40, no. 2, pp. 144–7, Feb., 1998.
p	Hatch, 1998	Children had an increased risk of acute lymphoblastic leukemia if they used electric blankets and other electrical appliances, or their mothers used electric blanket during pregnancy.	Hatch, E.E. et al., 'Association Between Childhood Acute Lymphoblastic Leukemia and Use of Electrical Appliances During Pregnancy and Childhood', *Epidemiology*, vol. 9, no. 3, pp. 234–45, May, 1998.
p	Wartenberg, 1998	Analysis of 15 studies found that there was a 'consistent risk' of childhood leukemia from residential magnetic fields.	Wartenberg, D., 'Residential Magnetic Fields and Childhood Leukemia: a Meta-analysis', *Am. J. Public Health*, vol. 88, no. 12, pp. 1787–94, Dec., 1998.
p	Green, 1999	Children exposed to high magnetic fields had an increased risk of leukemia; most exposed had a risk 4.5 times higher; children under six exposed to over 1.4 mG had 5.7 times the risk.	Green, L. M. et al., 'A Case-control Study of Childhood Leukemia in Southern Ontario, Canada, and Exposure to Magnetic Fields in Residences', *Int. J. Cancer*, vol. 82, no. 2, pp. 161–70, 19 Jul. 1999.
p	Angelillo, 1999	A meta-analysis of 14 studies found an increased risk of childhood leukemia based on wiring-configurations (almost 1.5 times the average risk) or measurements (over 1.5 times the average risk).	Angelillo, I.F. and Villari, P., 'Residential Exposure to Electromagnetic Fields and Childhood Leukemia: a Meta-analysis', *Bulletin of the World Health Organisation*, 1999, 77, pp. 906–14.
p	Villeneuve, 2000	Workers exposed to 10-40 V/m had 8–10 times the risk of	Villeneuve, P.J. et al., 'Leukemia in Electric Utility Workers: the

		leukemia.	Evaluation of Alternative Indices of Exposure to 60Hz Electric and Magnetic Fields', *Am. J. Ind. Med.,* vol. 37, no. 6, pp. 607–17, Jun., 2000.
p	Greenland, 2000	A review of 12 studies found that children exposed to under 3 mG had little risk of leukemia, but those exposed to over 3 mG had 1.7 times the average.	Greenland, S. et al., 'A Pooled Analysis of Magnetic Fields, Wire Codes and Childhood Leukemia', *Epidemiology,* vol. 11, no. 6, pp. 624–334, Nov., 2000.
p	Ahlbom, 2000	A review of nine studies found that children exposed to over 4 mG had double the average rate of leukemia.	Ahlbom, A. et al., 'A Pooled Analysis of Magnetic Fields and Childhood Leukemia', *British Journal of Cancer,* vol. 83, no. 5, pp. 692–8, Sep., 2000.
p	Doll, 2001	A review of studies found that children exposed to over 4 mG had double the average rate of leukemia.	NRPB, 'ELF Electromagnetic Fields and the Risk of Cancer', *Documents of the NRPB,* vol. 12, no. 1, 2001.
p	Schüz, 2001	Children exposed to more than 2 mG at night had a significantly increased risk of leukemia.	Schüz, J. et al., 'Residential Magnetic Fields as a Risk Factor for Childhood Acute Leukemia: Results from a German Population-based Case-control Study', *Int. J. of Cancer,* vol. 91, no. 5, pp. 728–35, March, 2001.
p rf	Milham, 1985	Workers in occupations assessed as having high electric or magnetic fields had greater mortality rates from leukemias and other lymphomas.	Milham, S., 'Mortality in Workers Exposed to Electromagnetic Fields', *Environ. Health Perspect.,* 62, pp. 297–300, Oct., 1985.
rf	Milham, 1988	Amateur radio operators with RF exposure had a higher rate of death from leukemias.	Milham, S. Jr., 'Increased Mortality in Amateur Radio Operators Due to Lymphatic and Hematopoietic Malignancies', *Am. J. Epidemiol.,* vol. 127, no. 1, pp. 50–4, Jan., 1988.
rf	Maskarinec, 1994	A cluster of children with leukemia in Hawaii was more likely to have lived within	Maskarinec, G. et al., 'Investigation of Increased Incidence in Childhood Leukemia

		2.6 miles of radio towers.	Near Radio Towers in Hawaii: Preliminary Observations', J. Environ. Pathol. *Toxicol. Oncol.,* vol. 13, no. 1, pp. 33-7, 1994.
rf	Hocking, 1996	Among people living near three TV towers in North Sydney there was a higher rate of childhood lymphatic leukemia.	Hocking, B. et al., 'Cancer Incidence and Mortality and Proximity to TV Towers', *Med. J. Aust.,* vol. 165, nos. 11–12, pp. 601–5, Dec., 1996.
rf	Dolk, 1997 TV, FM radio	People living near a TV tower and FM radio transmitter had a higher rate of adult leukemia. (Not supported by a second study.)	Dolk, H. et al., 'Cancer Incidence Near Radio and Television Transmitters in Great Britain. 1. Sutton Coldfield Transmitter', *Am. J. Epidemiol.,* vol. 145, no. 1, pp. 1-9, Jan. 1, 1997.
rf	Michelozzi, 1998	People living within a 3.5 km radius of a high-power radio-transmitter in Rome had a higher rate of leukemia, and risk for men declined with distance from the transmitter.	Michelozzi, P. et al., 'Risk of Leukemia and Residence Near a Radio Transmitter in Italy', *Epidemiology,* vol. 9 (Suppl.), pp. 354, 1998.
rf	Hocking, 2000	Children with acute lymphatic leukemia who lived nearest to the TV towers had decreased survival rates.	Hocking, B. and Gordon, I., 'Decreased Survival for Childhood Leukemia in Proximity to TV Towers', Annual Scientific Meeting of Royal Australasian College of Physicians in Adelaide, SA, 2–5 May 2000.

Breast cancer

p	Vena, 1991	Women who used electric blankets all night frequently had a slightly increased risk of breast cancer.	Vena, J. E. et al., 'Use of Electric Blankets and Risk of Postmenopausal Breast Cancer', *Am. J. Epidemiol.*, vol. 134, no. 2, pp. 180–5, Jul. 15, 1991.
p	Coogan, 1996	Women with potential for high EMR exposure had a slightly increased risk of breast cancer.	Coogan, P. F. et al., 'Occupational Exposure to 60-hertz Magnetic Fields and Risk of Breast Cancer in Women', *Epidemiology*, vol. 7, no. 5, pp. 459–64, Sep., 1996.
p	Feychting, 1998	Women under 50 with high exposures had a greater risk of breast cancer.	Feychting, M. et al., 'Occupational and Residential Magnetic Field Exposure and Female Breast Cancer', *EMF Research Review*, DOE, abstract-28, Sep., 1998.
p	Blackman, 2001	Breast cancer cells exposed to EMR and melatonin did not show the expected reduction in proliferation.	Blackman, C.F. et al., 'The Influence of 1.2 MicroT, 60 Hz Magnetic Fields on Melatonin- and Tamoxifen-induced Inhibition of MCF-7 Cell Growth', *Bioelectromagnetics,* vol. 22, no. 2, pp. 122–8, 2001.
p	Pollan, 2001	Men exposed to EMR at work had an increased risk of breast cancer.	Pollan, M. et al., 'Breast Cancer, Occupation, and Exposure to Electromagnetic Fields Among Swedish Men', *Am. J. Ind. Med.*, vol. 39, no. 3, pp. 276–85, 2001.
p+rf	Demers, 1991	Men with breast cancer were more likely to have electrical occupations.	Demers, P.A. et al., 'Occupational Exposure to Electromagnetic Fields and Breast Cancer in Men', *Am. J. Epidemiol.*, vol. 134, no. 4, pp. 340–7, Aug. 15, 1991.
p+rf	Loomis, 1994	Women working in electrical occupations had an increased incidence of mortality from breast cancer compared to other occupations.	Loomis, D.P. et al., 'Breast Cancer Mortality Among Female Electrical Workers in the United States', *J. Natl. Cancer Inst.*, vol. 86, no. 12, pp. 921–5, 15 Jun. 1994.

| p+rf | Pollan, 1999 | People in some occupations with higher breast cancer risk are exposed to higher levels of EMR. | Pollan, M. and Gustavsson, P., 'High-risk Occupations for Breast Cancer in the Swedish Female Working Population', *Am. J. Public Health*, vol. 89, no. 6, pp. 875–81, 1999. |
| rf | Tynes, 1996 | Women working as radio and telegraph operators had an increased rate of breast cancer. | Tynes, T. et al., 'Incidence of Breast Cancer in Norwegian Female Radio and Telegraph Operators', *Cancer Causes Control*, vol. 7, no. 2, pp. 197–204, Mar., 1996. |

Other cancers

p	Wertheimer & Leeper, 1982	Adults exposed to high-current electrical wiring configurations had a greater risk of cancer.	Wertheimer, N. and Leeper, E., 'Adult Cancer Related to Electric Wires Near the Home', *Int. J. Epidemiol.* vol. 11, no. 4, pp. 345–55, Dec. 1982.
p	Phillips, 1986	Colon cells exposed to electric and/or magnetic fields had a higher number of antigens which are associated with cancer.	Phillips, J.L. et al., 'In Vitro Exposure to Electromagnetic Fields: Changes in Tumour Cell Properties', *Int. J. Radiat. Biol. Relat. Stud. Phys. Chem. Med.*, vol. 49, no. 3, pp. 463–9, Mar., 1986.
p	Tomenius, 1986	Children exposed to 3 mG at home had a greater risk of cancer.	Tomenius, L., 50-Hz 'Electromagnetic Environment and the Incidence of Childhood Tumors in Stockholm County', *Bioelectromagnetics*, vol. 7, no. 2, pp. 191–207, 1986.
p	Savitz, 1988	Children with cancers were more likely to live in a home with a wiring configuration suggestive of higher magnetic fields.	Savitz, D.A., et al., 'Case-control Study of Childhood Cancer and Exposure to 60-Hz Magnetic Fields', *Am. J. Epidemiol.*, vol. 128, no. 1, pp. 21–38, Jul., 1998.

p	Olsen, 1993	A significant association was seen between all major types of childhood cancer combined and exposure to magnetic fields from high-voltage installations of > or = 0.4 micro T'(4mG).	Olsen, J.H. et al., 'Residence Near High Voltage Facilities and Risk of Cancer in Children', *Brit. Med. J.*, vol. 307, no. 6909, pp. 891–5, Oct. 9, 1993.
p	Floderus, 1994	Railway workers had an elevated risk of a range of cancers.	Floderus, B. et al., 'Incidence of Selected Cancers in Swedish Railway Workers, 1961-79', *Cancer Causes Control*, vol. 5, no. 2, pp. 189–94, 1994.
p	Uckun, 1995	Immune cells exposed to 60Hz fields activated enzyme protein tyrosine kinase which initiates a process that allows unchecked proliferation of cells — a hallmark of cancer.	Uckun, F.M. et al., 'Exposure of B-lineage Lymphoid Cells to Low Energy Electromagnetic Fields Stimulates Lyn Kinase', *J. Biol. Chem.*, vol. 270, no. 46, pp. 27666–70, Nov. 17, 1995.
p	Wertheimer, 1995	Children living in homes where plumbing produced high currents had an increased risk of cancer.	Wertheimer, N. et al., 'Childhood Cancer in Relation to Indicators of Magnetic Fields from Ground Current Sources', *Bioelectromagnetics*, vol. 16, no. 2, pp. 86–96, 1995.
p	Ji, 1999	Electricians had more than 7 times the risk of developing pancreatic cancer.	Ji, B.T. et al., 'Occupation and Pancreatic Cancer Risk in Shanghai, China', *Am. J. Ind. Med.*, vol. 35, no. 1, pp. 76–81, Jan., 1999.
p	Villeneuve, 2000	Workers in an electric utility who spent most time exposed to electric fields between 10-40 V/m had more than three times the rate of Non-Hodgkin's Lymphoma.	Villeneuve, P.J. et al., 'Non-Hodgkin's Lymphoma Among Electric Utility Workers in Ontario: the Evaluation of Alternate Indices of Exposure to 60Hz Electric and Magnetic Fields', *Occup. Environ. Med.*, vol. 57, no. 4, pp. 249–57, Apr., 2000.

rf	Byus, 1984	Human tonsil cells exposed to EMR at 450 MHz showed changes in activity of enzyme cAMP-protein kinase. (This can cause changes in cell membrane leading to cell proliferation, a hallmark of cancer.)	Byus, C.V. et al., 'Alterations in Protein Kinase Activity Following Exposure of Cultured Human Lymphocytes to Modulated Microwave Fields', *Bioelectromagnetics*, vol. 5, no. 3, pp. 341–51, 1984.
rf	Repacholi, 1997	Mice exposed to radiation of the type emitted by a mobile phone developed 2.4 the number of lymphomas as controls.	Repacholi, M.H. et. al., 'Lymphomas in E mu-Pim 1 Transgenic Mice Exposed to Pulsed 900 MHz Electromagnetic Fields', *Radiat. Res.*, vol. 147. no. 5. pp. 631–40, May, 1997.
rf	Lagorio, 1997	Workers exposed to EMR had a higher rate of death from malignancy neoplasms and an increased risk of leukemia.	Lagorio, S. et al., 'Mortality of Plastic-ware Workers Exposed to Radiofrequencies', *Bioelectromagnetics* vol. 18, no. 6, pp. 418–21, 1997.
rf	Finkelstein, 1998	Policemen working with radar guns had an increased risk of testicular cancer and melanoma.	Finkelstein, M.M., 'Cancer Incidence Among Ontario Police Officers', *Am. J. Ind. Med.*, vol. 34, no. 2, pp. 157–62, 1998.
rf mp	Stang, 2001	People who used mobile phones had four times the risk of developing eye cancer.	Stang, A. et al., 'The Possible Role of Radiofrequency Radiation in the Development of Uveal Melanoma', *Epidemiology,* vol. 12, no. 1, pp 7–12, Jan. 2001.

Heart problems

p	Sastre, 1998	Exposure to EMR of 200 mG affected heart-rate variability in the band associated with temperature and blood pressure control.	Sastre, A. et al., 'Nocturnal Exposure to Intermittent 60 Hz Magnetic Fields Alters Human Cardiac Rhythm', *Bioelectromagnetics*, vol. 19, no. 2, pp. 98–106, 1998.

p	Bortkiewicz, 1998	Electrical workers had an increased incidence of disturbances to electrocardiograph patterns.	Bortkiewicz, A. et al., 'Exposure to Electromagnetic Fields with Frequencies of 50 Hz and Changes in the Circulatory System in Workers at Electrical Power Stations', *Med. Pr.*, vol. 49, no. 3, pp. 261–74, 1998.
p	Savitz, 1999	People in jobs with high magnetic fields had an increased risk of death from arrhythmia-related conditions and acute myocardial infarction.	Savitz, D.A. et al., 'Magnetic Field Exposure and Cardiovascular Disease Mortality Among Electric Utility Workers', *Am. J. Epidemiol.*, vol. 149, no. 2, pp. 135–42, 15 Jan. 1999.
p	Sait, 1999	Volunteers exposed to occupational levels of EMR showed changes in heart rate.	Sait, M.L. et al., 'A Study of Heart Rate and Heart Rate Variability in Human Subjects Exposed to … Magnetic Fields', *Med. Eng. Phys.*, vol. 21, no. 5, pp. 361–9, 1999.
p	Ventura, 2000	Ventricular myocyte cells exposed to pulsed magnetic fields exhibited enhanced expression and transcription of opioid genes, which may affect the cardiovascular system.	Ventura, C. et al., 'Elf-pulsed Magnetic Fields Modulate Opioid Peptide Gene Expression in Myocardial Cells', *Cardiovasc. Res.*, vol. 45, no. 4, pp. 1054–64, Mar., 2000.
rf	Bortkiewicz, 1996 (broadcast stations)	Workers exposed to EMR had problems regulating cardiovascular function.	Bortkiewicz, A. et al., 'Heart Rate Variability in Workers Exposed to Medium-frequency Electromagnetic Fields', *J. Auton. Nerv. Syst.*, vol. 59, no. 3, pp. 91–7, Jul. 5, 1996.
rf	Bortkiewicz, 1997 (AM broadcast stations)	Workers with high EMR exposure had more frequent abnormalities in ECG.	Bortkiewicz, A. et al., 'Ambulatory ECG Monitoring in Workers Exposed to Electromagnetic Fields', *J. Med. Eng. Technol.*, vol. 21, no. 2, pp. 41–6, Mar.–Apr., 1997.
rf	Szmigielski, 1998	Workers exposed to EMR showed changes to diurnal rhythms of blood pressure and	Szmigielski, S. et al., 'Alteration of Diurnal Rhythms of Blood Pressure and Heart Rate to

		heart rate.	Workers Exposed to Radiofrequency Electromagnetic Fields', *Blood Press Monit.*, vol. 3, no. 6, pp. 323–30, 1998.
rf	Lu, 1999	Rats exposed to EMR had significantly decreased arterial blood pressure (hypotension).	Lu, S.T. et al., 'Ultrawide-band Electromagnetic Pulses Induced Hypotension in Rats', *Physiol. Behav.* vol. 67, no. 3, pp. 753–61, 1999.

Miscarriages, birth defects, and reproductive problems

elf/rf	Delgado, 1982	Chicken embryos exposed to EMR at 10, 100 and 1,000 Hz produced deformed foetuses at I mG.	Delgado, J.M.R. et al., *Journal of Anatomy*, vol. 134, pp. 533, 1982.
p	Nordstrom, 1983	Fathers who were high-voltage switchyard workers produced offspring with more congenital malformations.	Nordstrom, S. et al., 'Reproductive Hazards Among Workers at High Voltage Substations', *Bioelectromagnetics*, vol. 4, no. 1, pp. 91–101, 1983.
p	Spitz, 1985	Children whose fathers worked in occupations with high EMR exposure had an increased risk of death from neuroblastoma.	Spitz, M.R. and Johnson, C.C., 'Neuroblastoma and Paternal Occupation. A Case-control Analysis', *Am. J. Epidemiol*, vol. 121, no. 6, pp. 924–9, Jun., 1985.
p	Wertheimer, 1986	Users of electric blankets and heated waterbeds showed a trend for seasonal miscarriages.	Wertheimer, N. and Leeper, E., 'Possible Effects of Electric Blankets and Heated Waterbeds on Fetal Development', *Bioelectromagnetics*, vol. 7, no. 1, pp. 13–22, 1986.
p	Wertheimer, 1989	Increased risk of miscarriage occurred during seasons when EMR was increased.	Wertheimer, N. and Leeper, E., 'Fetal Loss Associated With Two Seasonal Sources of Electromagnetic Field Exposure', *Am. J. Epidemiol.*, vol. 129, no. 1, pp. 220–4, Jan., 1989.

p	Johnson and Spitz, 1989	Parents exposed to EMR and solvents produced children with a higher rate of death from brain tumours.	Johnson, C.C. and Spitz, M.R., 'Childhood Nervous System Tumours: an Assessment of Risk Associated with Paternal Occupations Involving Use, Repair or Manufacture of Electrical and Electronic Equipment', *Int. J. Epidemiol.*, vol. 18, no. 4, pp. 756–62, Dec., 1989.
p	Juutilainen, 1993	Women who had suffered pregnancy loss were more likely to have lived in a home with high EMR.	Juutilainen, J., et al., 'Early Pregnancy Loss and Exposure to 50-Hz Magnetic Fields', *Bioelectromagnetics,* vol. 14, no. 3, pp. 229–36, 1993
p	Li, 1995	Subfertile women who used electric blankets during pregnancy had 4.4 times the risk of producing offspring with congenital urinary tract anomalies.	Li, D.K. et al., 'Electric Blanket Use During Pregnancy in Relation to the Risk of Congenital Urinary Tract Anomalies Among Women With a History of Subfertility', *Epidemiology*, vol. 6, no. 5, pp. 485–9, Sep., 1995.
p	Stenlund, 1997	Males exposed to high magnetic fields at work had an increased risk of testicular cancer.	Stenlund, C. and Floderus, B., 'Occupational Exposure to Magnetic Fields in Relation to Male Breast Cancer and Testicular Cancer: a Swedish Case-control Study', *Cancer Causes Control*, vol. 8, no. 2, pp. 184–91, Mar., 1997.
p	Belanger, 1998	Women who used electric blankets at time of conception and early pregnancy had a slight increase in miscarriage rate.	Belanger, K. et al., 'Spontaneous Abortion and Exposure to Electric Blankets and Heated Water Beds', *Epidemiology*, vol. 9, no. 1, pp. 36–42, Jan. 1998.
p	Tornqvist, 1998	Fathers exposed to high fields at time of conception produced offspring with slightly higher proportion of malformations.	Tornqvist, S., 'Paternal Work in the Power Industry: Effects on Children at Delivery', *J. Occup. Environ. Med.,* vol. 40, no. 2, pp. 111–7, Feb., 1998.

p	Blaasaas, 1999	Parents exposed to high EMR had children with lower birth weight, shorter gestational age and higher death rates than those in low-exposure group.	Blaasaas, K.G. et al., 'Maternal and Paternal Occupational Exposure to 50 Hz Magnetic Fields and Pregnancy Outcomes. A Population Based Study', *BEMS* 1999, abstract 12-5.
p	Cecconi, 2000	Follicles from mouse ovaries exposed to 33 and 50 Hz fields failed to develop appropriately for reproduction.	Cecconi, S. et al., 'Evaluation of the Effects of Extremely Low Frequency Electromagnetic Fields on Mammalian Follicle Development', *Hum. Reprod.*, vol. 15, no. 11, pp. 2319–25, Nov. 2000.
p	Li, 2001	Women exposed to a maximum magnetic field of 16 mG or more had a greater risk of miscarriage.	*Microwave News*, May-Jun. 2001.
vdu	Goldhaber, 1988	Women who used VDUs for more than 20 hrs per week during first trimester of pregnancy had nearly double the risk of miscarriages	Goldhaber, M. K. et al., 'The Risk of Miscarriage and Birth Defects Among Women Who Use Visual Display Terminals During Pregnancy', *Am. J. Ind. Med.*, vol. 13, no. 6, pp. 695–706, 1988
vdu	Windham, 1990	In women with greater VDU use there was a slightly greater risk of intrauterine growth retardation.	Windham, G.C., 'Use of Video Display Terminals During Pregnancy and the Risk of Spontaneous Abortion, Low Birthweight, or Intrauterine Growth Retardation', *Am. J. Ind. Med.*, vol. 18, no. 6, pp. 675–88, 1990.
vdu	Lindbohm, 1992	Women who used VDUs with ELF fields over 9 mG had three and a half times the rate of spontaneous abortions.	Lindbohm, M.L. et al., 'Magnetic Fields of Video Display Terminals and Spontaneous Abortion', *Am. J. Epidemiol.*, vol. 136, no. 9, pp. 1041–51, Nov. 1992.
vdu	McDiarmid, 1994	People working in an office environment had an excess of spontaneous abortions (1.5–2.5 times the average).	McDiarmid, M.A. et al., 'Investigation of a Spontaneous Abortion Cluster: Lessons Learned', *Am. J. Ind. Med.*, vol. 25, no. 4., pp. 463–75, Apr., 1994.

vdu	Youbicier-Simo, 1997	In chicken embryos and young chickens exposed to continuous EMR from VDUs there was a 47–68 per cent increased miscarriage rate and depressed levels of several antibodies.	Youbicier-Simo, B.J., 'Biological Effects of Continuous Exposure of Embryos and Young Chickens to Electromagnetic Fields Emitted by Video Display Units', *Bioelectromagnetics,* vol. 18, no. 7, pp. 514–23, 1997.
rf	Goldsmith, 1995	People in military, broadcasting or other exposed occupations had a higher rate of adverse reproductive outcomes, especially increased spontaneous abortions.	Goldsmith, J.R., 'Epidemiologic Evidence of Radiofrequency Radiation (Microwave) Effects on Health in Military, Broadcasting and Occupational Studies', *Int. J. Occup. Environ. Health*, vol. 1, no. 1, pp. 47–57, Jan., 1995.
rf	Weyandt, 1996	US servicemen with potential microwave exposures had lower sperm counts.	Weyandt, T.B. et al., 'Semen Analysis of Military Personnel Associated with Military Duty Assignments', *Reprod. Toxicol.*, vol. 10, no. 6, pp. 521–8, Nov.–Dec., 1996.
rf	Magras, 1997	Mice were located near an antenna farm and mated. 'A progressive decrease in the number of newborns per dam was observed, which ended in irreversible infertility.'	Magras, I.N. and Xenos, T.D., 'RF Radiation-induced Changes in the Prenatal Development of Mice', *Bioelectromagnetics,* vol. 18, no. 6, pp. 455–61, 1997.
rf mp	Youbicier-Simo, 1998	In chicken eggs exposed to EMR from mobile phones more embryos died.	Youbicier-Simo, G.J. et al., 'Mortality of Chicken Embryos Exposed to EMFs From Mobile Phones', 20th Annual Meeting of the Bioelectromagnetics Society, St. Pete Beach, FL, June, 1998.
rf	Dasdag, 1999	Rats exposed to EMR from mobile phones in speech mode showed changes in testes.	Dasdag, S. et al., 'Whole-body Microwave Exposure Emitted by Cellular Phones and Testicular Function of Rats', *Urol. Res.*, vol. 27, no. 3, pp. 219–23, 1999.
rf	Afromeev, 1999	Rats exposed to EMR showed testicular differences.	Afromeev, V.I. and Tkachenko, V.N., 'Change in the Percent of Lactate Dehydrogenase

Isoenzyme Level in Testes of
Animals Exposed to Superhigh
Frequency Radiation', *Biofizika,*
vol. 44, no. 5, pp. 931–2, 1999.

Changes in electrical activity of brain

dc	von Klitzing, 1989	A static magnetic field induced a change of brain function shown on EEG.	von Klitzing, L., 'Static Magnetic Fields Increase the Power Intensity of EEG of Man', *Brain Res.*, vol. 483, no. 1, pp. 201–3, 27 Mar. 1989.
vlf	Marino, 1996	Subjects exposed to 1.5 and 10 Hz bursts of EMR exhibited altered electrical activity of the brain.	Marino, A.A. et al., 'Low-level EMFs are Transduced Like Other Stimuli', *J. Neurol. Sci.*, vol. 144, nos. 1–2, pp. 99–106, Dec., 1996.
vlf	Bell, 1994a	Subjects exposed to fields of 10 Hz at 1 G for 10 minutes showed significantly reduced electrical activity of the brain after the exposure ceased.	Bell, G.B. et al., 'Frequency-specific Blocking in the Human Brain Caused by Electromagnetic Fields', *Neuroreport,* vol. 5, no. 4, pp. 510–2, 12 Jan. 1994.
vlf	Bell, 1994b	Subjects exposed to 1.5 and 10-Hz bursts of EMR exhibited altered electrical activity of the brain.	Bell, G.B. et al., 'Frequency-specific Responses in the Human Brain Caused by Electromagnetic Fields', *J. Neurol. Sci.*, vol. 123, nos. 1–2, pp. 26–32, May, 1994.
vlf	Bise, 1998	Volunteers exposed to 680 mG between 3 and 75 Hz showed changes to EEG patterns.	Bise, W. L. et al., 'Multiple Extremely Low Frequency Magnetic and Electromagnetic Field Effects on Human Electroencephalogram and Behavior', EPRI, Annual Review of Research on Biological Effects of Electric and Magnetic Fields from the Generation, Delivery and Use of Electricity, Sept., 1998.
p	Bell, 1992	Human subjects exposed to static and 60 Hz magnetic fields exhibited changes to EEG.	Bell, G.B. et al., 'Alterations in Brain Electrical Activity Caused by Magnetic Fields: Detecting

			the Detection Process', Electroencephalogr. Clin. *Neurophysiol.*, vol. 83, no. 6, pp. 389–97, Dec., 1992.
rf	Bise, 1978	People exposed to EMR at below urban levels showed changes in brain waves and behaviour.	Bise, W., 'Low Power Radio-frequency and Microwave Effects on Human Electroencephalogram and Behavior', *Physiol. Chem. Phys.*, vol. 10, no. 5, pp. 387–98, 1978.
rf	Shandala, 1979	Animals exposed to microwaves showed changes in electrical activity of brain and other parameters.	Shandala, M.G. et al., 'Study of Nonionizing Microwave Radiation Effects Upon the Central Nervous System and Behavior Reactions', *Environ. Health Perspect.*, vol. 30, no. 2, pp. 115–21, Jun., 1979.
rf mp	von Klitzing, 1995	People exposed to pulsed EMR from mobile phones exhibited changes in EEG especially in alpha range and effects lasted for some hours after exposure.	von Klitzing, L., 'Low-frequency Pulsed Electromagnetic Fields Influence EEG of Man', *Phys. Medica,* vol. 11, pp. 77–80, 1995.
rf mp	Reiser, 1995	People exposed to signals from a digital mobile phone showed increased in EEG power in several frequency bands c. 27–32, 15 minutes after exposure.	Reiser, H. et al., 'The Influence of Electromagnetic Fields on Human Brain Activity', *Eur. J. Med. Res.* ,vol. 16, no. 1, pp. Oct., 1995.
rf mp	Mann & Roschke, 1996	Subjects exposed to high-frequency fields had sleep disturbance and altered EEG.	Mann, K. and Roschke, J., 'Effects of Pulsed High-frequency Electromagnetic Fields on Human Sleep', *Neuropsychobiology,* vol. 33, no. 1, pp. 41–7, 1996.
rf mp	Eulitz, 1998	Pulsed radiation from mobile phones affected the brain's electrical response.	Eulitz, C. et al., 'Mobile Phones Modulate Response Patterns of Human Brain Activity', *Neuroreport,* vol. 9, no. 14, pp. 3229–3232, Oct. 5,1998.

rf mp	Borbely, 1999	Volunteers exposed to radiation from mobile phones in alternating pattern of 15 minutes on/off showed significantly altered EEG and some improvement in sleep.	Borbely, A. A. et al., 'Pulsed High-frequency Electromagnetic Field Affects Human Sleep and Sleep Electroencephalogram', *Neurosci. Lett.*, vol. 275, no. 3, pp. 207–10, Nov. 19, 1999.
rf mp	Freude, 2000	Volunteers exposed to radiation from a mobile phone showed a significant decrease in brain waves known as slow brain potential.	Freude, G. et al., 'Microwaves Emitted by Cellular Telephones Affect Human Slow Brain Potentials', *Eur. J. Appl. Physiol.*, vol. 81, nos. 1–2, pp. 18–27, Jan., 2000.
rf mp	Huber, 2000	Volunteers exposed to radiation of 900 MHz from mobile phones for 30 minutes just before sleep showed changes to sleep. 'Thus the changes of brain function induced by pulsed high-frequency EMF outlast the exposure period.'	Huber, R. et al., 'Exposure to Pulsed High-frequency Electromagnetic Field During Waking Affects Human Sleep EEG', *Neuroreport,* vol. 11, no. 15, pp. 3321–5, Oct., 2000
rf mp	Lebedeva, 2000	Volunteers exposed to radiation from a mobile phone (902.4 MHz at 0.06 mW/cm2) for 15 mins showed changes to EEG activity indicating 'brain functional state changes'.	Lebedeva, N.N. et al., 'Cellular Phone Electromagnetic Field Effects on Bioelectric Activity of Human Brain', *Crit. Rev. Biomed. Eng.*, vol. 28, no. 1–2, pp. 323–37, 2000.
rf	Krause, 2000	People exposed to mobile phone radiation showed changes to EEG during memory task.	Krause, C.M. et al., 'Effects of Electromagnetic Field Emitted by Cellular Phones on the EEG During a Memory Task', *Neuroreport,* vol. 11, no. 4, pp. 761–4, 20 Mar. 2000.

Effects on the nervous system

p	Reichmanis, 1979	There was a correlation between fields from powerlines and occurrence of suicide.	Reichmanis, M. et al., 'Relation Between Suicide and the Electromagnetic Field of Overhead Power Lines', *Physiol. Chem. Phys.*, vol. 11, no. 5, pp. 395–403, 1979.

p	Perry, 1981	People living near powerlines had an increased risk of suicide.	Perry, F.S. et al., 'Environmental Power-frequency Magnetic Fields and Suicide', *Health Phys.*, vol. 41, no. 2, pp. 267–77, Aug., 1981.
p	Wolpaw, 1987	Monkeys exposed to powerline fields for 3 weeks had depressed levels of dopamine & serotonin which are related to depression and behavioural changes.	Wolpaw, J., 'Biological Effects of Power Line Fields', Albany, N.Y., New York State Power-Lines Project Scientific Advisory Panel, 1987.
p	Perry & Pearl, 1988	People living in blocks of flats with underfloor or electric storage heating systems had an increased rate of depression.	Perry, S. and Pearl, L., 'Power Frequency Magnetic Field and Illness in Multi-storey Blocks', *Public Health*, vol. 102, no. 1, pp. 11–18, Jan., 1988.
p	Perry, 1989	People with depressive illnesses had higher levels of EMR outside their homes.	Perry, S. et al., 'Power Frequency Magnetic Field; Depressive Illness and Myocardial Infarction', *Public Health*, vol. 103, no. 3, pp. 177–80, May, 1989.
p	Poole, 1993	People who lived near powerlines reported a higher incidence of depressive symptoms.	Poole, C. et al., Depressive 'Symptoms and Headaches in Relation to Proximity of Residence to an Alternating-current Transmission Line Right-of-way', *Am. J. Epidemiol.*, vol. 137, no. 3, pp. 318–30, 1 Feb. 1993.
p	Baris, 1996	More exposed groups of electrical utility workers, Quebec, had a small increase in suicide rate. There was some evidence of a cumulative effect.	Baris, D. et al., 'A Case Cohort Study of Suicide in Relation to Exposure to Electric and Magnetic Fields Among Electrical Utility Workers', *Occupation. Environ. Med.*, vol. 53, no. 1, pp. 17–24, Jan., 1996.
p	Beale, 1997	People exposed to high-voltage powerlines showed significant psychological and mental health effects.	Beale, I.L., et al., 'Psychological Effects of Chronic Exposure to 50 Hz Magnetic Fields in Humans Living Near Extra-high-voltage

			Transmission Lines', *Bioelectromagnetics*, vol. 18, no. 8, pp. 584–94, 1997.
p	Verkasalo, 1997	Subjects living within 100 m of a high–voltage power line had 4.7 times the risk of severe depression.	Verkasalo, P.K. et al., 'Magnetic Fields of Transmission Lines and Depression', *Am. J. Epidemiol.*, vol. 146, no. 12, pp. 1037–45, 15 Dec. 1997.
p	Bonhomme-Faivre, 1998	People exposed to EMR had symptoms of fatigue, psychical asthenia, lipothymia, decreased libido, melancholy, depression and irritability.	Bonhomme-Faivre, L. et al., 'Study of Human Neurovegetative and Hematologic Effects of Environmental Low-frequency (50-Hz) Electromagnetic Fields Produced by Transformers', *Arch. Environ. Health*, vol. 53, no. 2, pp. 87–92, Mar.–Apr., 1998.
p	van Wijngaarden, 2000	Electricians working for five US electrical utilities had double and linesmen had 1.5 times the normal rate of suicide, and risk grew as exposure increased.	Van Wijngaarden, E. et al., 'Exposure to Electromagnetic Fields and Suicide Among Electricity Utility Workers: A Nested Case Control Study', *Occupational and Environmental Medicine*, vol. 57, 258–63, 15 Mar. 2000.
rf rad	Alpeter, 1995	People living near radar showed nervousness, restlessness, difficulty falling asleep and maintaining sleep, joint pains, changes to psychovegetative index, concentration problems, weakness, tiredness, constipation, diarrhoea, lower back pain and poor scholastic performance	Alpeter, E.S. et al., *Study of Health Effects of Short-wave Transmitter Station of Schwarzenburg, Berne, Switzerland*, University of Berne, Institute for Social and Preventative Medicine, Aug., 1995.
rf mp	Mann & Roschke, 1996	Subjects exposed to high-frequency fields had sleep disturbance and altered EEG.	Mann, K. and Roschke, J., 'Effects of Pulsed High-frequency Electromagnetic Fields on Human Sleep', *Neuropsychobiology*, vol. 33, no. 1, pp. 41–7, 1996.

rf mp	Mild, 1998	Mobile phone users showed symptoms of fatigue, headache, concentration problems, dizziness, discomfort, memory loss, warmth on/behind ear, burning skin, tingling, tightness.	Mild, K. J., 'Comparison of Symptoms Experienced by Users of Analogue and Digital Mobile Phones: a Swedish-Norwegian epidemiological study', Rept. No. 1998: 23, National Institute for Working Life, Sweden, 1998.
rf mp	Hocking, 1998	Users of mobile phones reported symptoms of burning or a dull ache in temporal, occipital or auricular areas.	Hocking, B., 'Preliminary Report: Symptoms Associated with Mobile Phone Use', *Occup. Med.* (Lond.), vol. 48, no. 6, pp. 357–60, Sep., 1998.
rf rad	Abelin, 1998	People living within 1.5 km of a radar reported more sleep problems, headaches, fatigue, irritability, low-back ache and limb pain, and children performed worse at school than those living over four km away.	Abelin, T. et al., 'Study on Health Effects of the Shortwave Transmitter Station of Schwarzenburg, Bern, Switzerland', Study No. 55, Aug., 1995, Swiss Federal Office of Energy.
rf	Marraccini, 1990	Foundry workers with prolonged exposures showed differences in neuropsychological performance.	Marraccini, P. et al., 'Evaluation of Neuropsychological Parameters in a Group of Metal Mechanics Occupationally Exposed to Radio Frequencies', *Med. Lav.*, vol. 81, no. 5, pp. 414–21, Sep.–Oct., 1990.
rf mp	Hladky, 1999	Volunteers exposed to EMR from a GSM mobile phone performed worse in a task requiring switching attention. This suggests risks of phone use while driving.	Hladky, A. et al., 'Acute Effects of Using a Mobile Phone on CNS Functions', *Cent. Eur. J. Public Health*, vol. 7, no. 4, pp. 165–7, 1999.
rf mp	Chia, 2000	Mobile phone users reported more headaches than non-users, and the prevalence of headaches increased with phone use.	Chia, S., 'Increase Prevalence of Headache Among Mobile Phone Users in Singapore—a Community Study', *Brit Med Jnl.*, vol. 321, no. 7245, 2000, letter.
rf mp	Oftedal, 2000	31% of Norwegian and 13% of Swedish survey respondents	Oftedal, G. et al., 'Symptoms Experienced in Connection with

		experienced symptoms from mobile phone use, including warmth on/near ear, burning on skin and headaches.	Mobile Phone Use', *Occup. Med.* (Lond.), vol. 50, no. 4, pp. 237–45, May, 2000.
rf mp	Hocking, 2000	A patient who used a mobile phone for long periods developed neurological symptoms on the scalp, adjacent to where the phone was held.	Hocking, B. and Westerman, R., 'Case Report: Neurological Abnormalities Associated With Mobile Phone Use', *Occup. Med.*, vol. 50, no. 5, pp. 366–8, 2000.

Neurodegenerative diseases

Alzheimer's Disease

p	Sobel, 1995	People working in occupations with exposure to EMR had three times the average risk of Alzheimer's disease.	Sobel, E. et al., 'Occupations with Exposure to Electromagnetic Fields: a Possible Risk Factor for Alzheimer's Disease', *Am. J. Epidemiol.*, vol. 142, no. 5, pp. 515–24, 1 Sep 1995.
p	Sobel, 1996	People working in occupations with high exposure to EMR had an increased risk of Alzheimer's disease.	Sobel, E. et al., 'Elevated Risk of Alzheimer's Disease Among Workers With Likely Electromagnetic Field Exposure', *Neurology*, vol. 47, no. 6, pp. 1477–81, Dec., 1996.
p	Sobel, 1996	EMR may cause Alzheimer's by stimulating amyloid beta.	Sobel, E. and Davanipour, Z., 'Electromagnetic Field Exposure May Cause Increased Production of Amyloid Beta and Eventually Lead to Alzheimer's Disease', *Neurology*, vol. 47, no. 6, pp. 1594–600, Dec., 1996.
p	Schulte, 1996	An excess of neurodegenerative diseases was found in occupations in several categories including people exposed to EMR.	Schulte, P.A., et al., 'Neurodegenerative Diseases: Occupational Occurrence and Potential Risk Factors, 1982 Through 1991', *Am. J. Public Health*, vol. 86, no. 9, pp. 1281–8, Sep., 1996.

| p | Feychting, 1998 | People in occupations exposed to high fields had 2.4 times the risk of Alzheimer's and 3.3 times the risk of dementia. | Feychting, M. et al., 'Dementia and Occupational Exposure to Magnetic Fields', *Scand. J. Work Environ. Health*, vol. 24, no. 1, pp. 46–53, Feb., 1998. |

Lou Gehrig's disease (Amyotrophic lateral sclerosis, ALS)

p	Davanipour, 1997	People with occupational exposure to EMR had an increased risk of ALS.	Davanipour, Z. et al., 'Amyotrophic Lateral Sclerosis and Occupational Exposure to Electromagnetic Fields', *Bioelectromagnetics*, vol. 18, no. 1, pp. 28–35, 1997.
p	Savitz, 1998	People working in exposed jobs in electric utilities had a higher rate of death from ALS.	Savitz, D. et al., 'Magnetic Field Exposure and Neurodegenerative Disease Mortality Among Electric Utility Workers', *Epidemiology*, vol. 9, no. 4, Jul., pp. 398–404, 1998.
p	Savitz, 1998	Electrical power plant operators had a risk of ALS 2 to five times the average.	Savitz, D.A. et al., 'Electrical Occupations and Neurodegenerative Disease: Analysis of US Mortality Data', *Arch. Environ. Health*, vol. 53, no. 1, pp. 71–4, Jan.–Feb., 1998.
p	Johansen, 1998	Danish electric utility workers had a rate of ALS twice the average.	Johansen, C. and Olsen, J.H., 'Mortality from Amyotrophic Lateral Sclerosis, Other Chronic Disorders, and Electric Shocks Among Utility Workers', *Am. J. Epidemiol.*, vol. 148, no. 4, pp. 362–8, Aug. 15, 1998.
p	Johansen, 2000	Workers at Danish electrical utilities had higher risks of senile dementia and motor neuron diseases.	Johansen, C., 'Exposure to Electromagnetic Fields and Risk of Central Nervous System Disease in Utility Workers', *Epidemiology*, vol. 11, no. 5, pp. 539–43, Sep., 2000.

Learning and performance

p	Salzinger, 1987	Rats exposed to powerline fields as foetuses later had problems learning and made more mistakes than unexposed rats.	Salzinger, K., 'Biological Effects of Power Line Fields', Albany, N.Y., New York State Power-Lines Project Scientific Advisory Panel, 1987.
p	Lai, 1996	Rats exposed to 60 Hz magnetic field performed worse in spatial memory task.	Lai, H., 'Spatial Learning Deficit in the Rat After Exposure to a 60 Hz Magnetic Field', *Bioelectromagnetics*, vol. 17, no. 6, pp. 494–6, 1996.
p	Kavaliers, 1996	Female mice exposed to brief (5 min.) 60 Hz magnetic field of 1000 mG showed improved learning.	Kavaliers, M. et al., 'Spatial Learning in Deer Mice: Sex Differences and the Effects of Endogenous Opioids and 60 Hz Magnetic Fields', *J. Comp. Physiol.* [A], vol. 179, no. 5., pp. 715–24, Nov., 1996.
p	Lai, 1998	Rats exposed to 60 Hz magnetic field for 1 hr performed worse in long-term memory test and used different strategies to find their way.	Lai, H. et al., 'Acute Exposure to a 60 Hz Magnetic Field Affects Rats' Water-maze Performance', *Bioelectromagnetics*, vol. 19, no. 2, pp. 117–22, 1998.
p	Sienkiewicz, 1998	Rats exposed to 50 Hz magnetic fields were less quick at learning a task.	Sienkiewicz, Z.J. et al., 'Deficits in Spatial Learning After Exposure of Mice to a 50 Hz Magnetic Field', *Bioelectromagnetics*, vol. 19, no. 2, pp. 79–84, 1998.
p	Trimmel, 1998	Humans exposed to 50 Hz fields showed reduced attention, perception and memory performance.	Trimmel, M. and Schweiger, E., 'Effects of an ELF (50 Hz, 1 mT) Electromagnetic Field (EMF) on Concentration in Visual Attention, Perception and Memory Including Effects of EMF Sensitivity', *Toxicol. Lett.*, 96–7, pp. 377–82, Aug., 1998.

rf	Lai, 1994	Rats exposed to microwaves at 2450 MHz showed deficit in spatial working memory.	Lai, H. et al., 'Microwave Irradiation Affects Radial-arm Maze Performance in the Rat', *Bioelectromagnetics*, vol. 15, no. 2, pp. 95–104, 1994.
rf rad	Kolodynski, 1996	Children living near a radar transmitter had impaired motor function, memory, attention, reaction times and neuromuscular apparatus endurance.	Kolodynski, A.A. and Kolodynska, V.V., 'Motor and Psychological Functions of School Children Living in the Area of the Skrunda Radio Location Station in Latvia', *Sci. Total Environ.*, vol. 180, no. 1, pp. 87–93, 2 Feb. 1996
rf	Grigor'ev, 1998	Among chicks exposed to low levels of EMR there was an effect on memory (imprinting).	Grigor'ev, IuG and Stepanov, V.S., 'Forming of Memory (Imprinting) in Chicks After Prior Low-level Exposure to Electromagnetic Fields', *Radiats. Biol. Radioecol.*, vol. 38, no. 2, pp. 223–31, 1998.
rf	Duan, 1998	Workers exposed to EMR performed worse in tests to assess neurobehavioural function than controls.	Duan, L. et al., 'Observations of Changes in Neurobehavioral Functions in Workers Exposed to High-frequency Radiation', *Chung Hua Yu Fang I Hsueh Tsa Chih* vol. 32, no. 2, pp. 109–11, 1998.
rf mp	Preece, 1999	Students exposed to mobile phone EMR improved on yes/no responses in performance tasks.	Preece, A.W. et al., 'Effect of a 915 MHz Simulated Mobile Phone Signal on Cognitive Function in Man', *Int. J. Radiat. Biol.*, vol. 75, no. 4, pp. 447–56, Apr., 1999.
rf	Wu, 1999	Rats exposed to EMR had more problems learning a maze task than controls.	Wu, J. et al., 'Influence of EMP on the Nervous System of Rats', *ACTA Biophysica Sinica,* vol. 15, pp. 152–7, 1999.
rf	Wang & Lai, 2000	Rats exposed to pulsed radiation suffered long-term memory loss.	Wang, B. and Lai, H., 'Acute Exposure to Pulsed 2450–MHz Microwaves Affects Water-maze Performance of Rats', *Bioelectromagnetics*, vol. 21, no. 1, pp. 52–6, Jan., 2000.

| rf | Koivisto, 2000 | Volunteers exposed to radiation from a mobile phone showed improved performance time at cognitive tasks. | Koivisto, M. et al., 'Effects of 902 MHz Electromagnetic Field Emitted by Cellular Telephones on Response Times in Humans', *Neuroreport*, vol. 11, no. 2, pp. 413–15, Feb., 2000. |

Sleep

p	Akerstedt, 1999	Volunteers exposed to a 50 Hz field of 10 mG had reductions in total sleep time, sleep efficiency, slow-wave sleep and slow-wave activity.	Akerstedt, T. et al., 'A 50-Hz Electromagnetic Field Impairs Sleep', *J. Sleep Res.*, vol. 8, no. 1, pp. 77–81, Mar., 1999.
p	Graham, 1999	Volunteers exposed to intermittent fields of 60 Hz, 283 mG, had less total sleep, reduced sleep efficiency, increased time in Stage II sleep and decreased REM sleep. They reported sleeping less well and feeling less rested.	Graham, C. and Cook, M.R., 'Human Sleep in 60 Hz Magnetic Fields', *Bioelectromagnetics*, vol. 20, no. 5, pp. 277–83, 1999.
rf mp	Mann, 1996	Volunteers exposed to pulsed mobile phone signal went to sleep more quickly, had less REM (rapid eye movement) sleep and showed changes to EEG during REM sleep.	Mann, K. and Roschke, J., 'Effects of Pulsed High-frequency Electromagnetic Fields on Human Sleep', *Neuropsychobiology*, vol. 33, no. 1, pp. 41–7, 1996.
rf	Borbely, 1999	Volunteers exposed to EMR from a GSM mobile phone showed changes to sleep (sleep was promoted), and brain pattern was modified during sleep.	Borbely, A.A. et al., 'Pulsed High-frequency Electromagnetic Field Affects Human Sleep and Sleep Electroencephalogram', *Neurosci. Lett.*, vol. 275, no. 3, pp. 207–10, Nov. 19, 1999.
rf mp	Huber, 2000	In volunteers exposed to radiation from a digital mobile phone for 30 minutes, the brain's electrical signals showed enhanced intensity in some frequencies during later sleep.	Huber, R. et al., 'Exposure to Pulsed High-frequency Electromagnetic Field During Waking Affects Human Sleep EEG', *Neuroreport,* vol. 11, no. 15, pp. 3321–5, 2000.

Effects on immune activity

p	Mevissen, 1996	Rats exposed to EMR exhibited depression of T cell proliferation.	Mevissen, M. et al., 'Exposure of DMBA-treated Female Rats in a 50-Hz, 50 microTesla Magnetic Field: Effects on Mammary Tumor Growth, Melatonin Levels, and T Lymphocyte Activation', *Carcinogenesis*, vol. 17, no. 5, pp. 903–10, May, 1996.
p	Uckun, 1995	Immune cells exposed to 60Hz fields activated enzyme protein tyrosine kinase which initiates a process that allows unchecked proliferation of cells—a hallmark of cancer.	Uckun, F.M. et al., 'Exposure of B-lineage Lymphoid Cells to Low Energy Electromagnetic Fields Stimulates Lyn Kinase', *J. Biol. Chem.*, vol. 270, no. 46, pp. 27666–70, 17 Nov. 1995.
p	Bonhomme-Faivre, 1998	People working near transformers and high-tension cables had a significant reduction in lymphocytes.	Bonhomme-Faivre, L. et al., 'Study of Human Neurovegetative and Hematologic Effects of Environmental Low-frequency (50-Hz) Electromagnetic Fields Produced by Transformers', *Arch. Environ. Health*, vol. 53, no. 2, pp. 87–92, Mar.–Apr., 1998.
rf	Lyle, 1983	Animals exposed to 450 Hz radiation developed temporary suppression of immune system.	Lyle, D.B. et al., 'Suppression of T-Lymphocyte Cytotoxicity Following Exposure to Sinusoidally Amplitude-modulated Fields', *Biolectricmagnetics*, vol. 4, no. 3, pp. 282–92, 1983.
rf	Shandala, 1983	Mice and rats exposed to 2375 MHz for three months had suppressed immunity.	Shandala, M.G. et al., 'Effect of Microwave Radiation on Cellular Immunity Indices in Conditions of Chronic Exposure', *Radiobiologiia*, vol. 23, no. 4, pp. 544–6, Jul.–Aug., 1983.
rf	Byus, 1984	Tonsil lymphocytes exposed to RF showed reduction in	Byus, C.V. et al., 'Alterations in Protein Kinase Activity Following

		lymphocyte protein kinase activity in certain windows of exposure.	Exposure of Cultured Human Lymphocytes to Modulated Microwave Fields', *Bioelectromagnetics*, vol. 5, no. 3, pp. 341–51, 1984.
rf	Cleary, 1990	In blood cells exposed to RF there were alterations to the proliferation of lymphocytes.	Cleary, S.F. et al., 'In Vitro Lymphocyte Proliferation Induced by Radio-frequency Electromagnetic Radiation Under Isothermal Conditions', *Bioelectromagnetics*, vol. 11, no. 1, pp. 47–56, 1990.
rf	Veyret, 1991	Mice exposed to amplitude - modulated waves showed a significant augmentation or weakening of immune responses, depending upon the frequency.	Veyret, B. et al., 'Antibody Responses of Mice Exposed to Low-power Microwaves Under Combined, Pulse-and-amplitude Modulation', *Bioelectromagnetics*, vol. 12, no. 1, pp. 47–56, 1991.
rf	Elekes, 1996	Male mice exposed to continuous or amplitude-modulated waves showed changes to immune response in spleen.	Elekes, E. et al., 'Effect on the Immune System of Mice Exposed Chronically to 50 Hz Amplitude Modulated 2.45 GHz Microwaves', *Bioelectromagnetics*, vol. 17, no. 3, pp. 246–8, 1996.
rf	Cleary, 1996	Immune cells exposed to EMR at 2450 MHz showed changes to proliferation.	Cleary, S.F. et al., 'Effect of Isothermal Radiofrequency Radiation on Cytolytic T Lymphocytes', *FASEB J.*, vol. 10, no. 8, pp. 913–9, Jun., 1996.
rf	Detlavs, 1996	Exposure of rats with wounds to EMR affected the healing process.	Detlavs, I. et al., 'Experimental Study of the Effects of Radiofrequency Electromagnetic Fields on Animals with Soft Tissue Wounds', *Sci. Total Environ.*, vol. 180, no. 1, pp. 35–42, 1996.
rf	Donnellan, 1997	Mast cells exposed to 835 MHz exhibited changes to cell growth,	Donnellan, M et al., 'Effects of Exposure to Electromagnetic

		gene expression and release of histamine.	Radiation at 835 MHz on Growth, Morphology and Secretory Characteristics of a Mast Cell Analogue, RBL-2H3', *Cell Biol. Int.*, vol. 21, no. 7, pp. 427–39, 1997.
rf	Repacholi, 1997	Mice exposed to radiation of the type emitted by a mobile phone developed 2.4 the number of lymphomas as controls.	Repacholi, M.H. et. al., 'Lymphomas in E mu-Pim 1 Transgenic Mice Exposed to Pulsed 900 MHz Electromagnetic Fields', *Radiat. Res.*, vol. 147. no. 5. pp. 631–40, May, 1997.
rf	Obukhan, 1998	Cells of the immune system exposed to EMR showed structural and functional changes.	Obukhan, K.I., 'The Effect of Ultrahigh-frequency Radiation on Adaptation Thresholds and the Damages to Blood System Cells', *Lik Sprava,* vol. 7, pp. 71–3, 1998.
rf	Garaj-Vrhovac, 1999	Workers exposed to EMR showed changes in immune cells.	Garaj-Vrhovac, V, 'Micronucleus Assay and Lymphocyte Mitotic Activity in Risk Assessment of Occupational Exposure to Microwave Radiation', *Chemosphere,* vol. 39, no. 13, pp. 2301–12, 1999.
rf	Trosic, 1999	Rats exposed to EMR showed decreased leukocytes and lymphocytes.	Trosic, I. et al., 'Animal Study on Electromagnetic Field Biological Potency', *Arh Hig Rada Toksikol.*, vol. 50, no. 1, pp. 5–11, 1999.

EMR affects hormones

vlf	Jacobson, 1994	People with neurological problems exposed to a very weak field near the pineal gland had increased melatonin production, and neurological performance improved over time.	Jacobson, J.I., 'Pineal-hypothalamic Tract Mediation of Picotesla Magnetic Fields in the Treatment of Neurological Disorders', *Panminerva Med.*, vol. 36, no. 4, pp. 201–5, Dec., 1994.
p	Wilson, 1990	Some people exposed to 60-Hz fields from electric blankets with	Wilson, B.W. et al., 'Evidence for an Effect of ELF Electromagnetic

		high fields showed changes to melatonin levels.	Fields on Human Pineal Gland Function', *J. Pineal Res.*, vol. 9, no. 4, pp. 259–69, 1990.
p	Beale, 1997	People exposed to EMR from powerlines had a greater incidence of type II diabetes.	Beale, I.L. et al., 'Psychological Effects of Chronic Exposure to 50Hz Magnetic Fields in Humans Living Near Extra-high-voltage Transmission Lines', *Bioelectromagnetics*, vol. 18, no. 8, pp. 584–94, 1997.
p	Wilson, 1999	Hamsters exposed to a 60 Hz field developed changes in hormones. 'One-time and repeated exposure to a 0.1 mT, 60 Hz MF can give rise to neuroendocrine responses in Phodopus.'	Wilson, B.W. et al., 'Effects of 60 Hz Magnetic Field Exposure on the Pineal and Hypothalamic-pituitary-gonadal axis in the Siberian Hamster (Phodopus Sungorus)', *Bioelectromagnetics*, vol. 220, no. 4, pp. 224–32, 1999.
p	Floderus, 1999	People with cancer exposed to high levels of EMR tended to develop cancer of the testes and uterus. 'The outcome suggests an interaction with the endocrine/immune system.'	Floderus, B. et al., 'Occupational Magnetic Field Exposure and Site-specific Cancer Incidence: A Swedish Cohort Study', *Cancer Causes Control*, vol. 10, no. 5, pp. 323–32, Oct. 1999.
p	Massot, 2000	Rats exposed to EMR showed changes in 5-HT(1B) receptors in brain (involved in mood disorders).	Massot, O. et al., 'Magnetic Field Desensitizes Receptor in Brain, Pharmacological and Functional Studies', *Brain Res.*, vol. 858, no. 1, pp. 143–50, Mar. 6, 2000.
rf	Navakatikian, 1994	Discusses low-intensity microwave exposure on stress hormones, insulin and sex hormones.	Navakatikian and Tomashevskaya, L.A., 'Phasic Behavioral and Endocrine Effects of Microwaves of Nonthermal Intensity in Biological Effects of Electric and Magnetic Fields', vol. 1, D.O. Carpenter, ed., Academic Press, San Diego, CA., pp. 333–42, 1994.
rf	Bielski, 1996	Workers exposed to radio waves showed problems with	Bielski, J. and Sikorski, M., 'Disturbance of Glucose

		sugar metabolism.	Tolerance in Workers Exposed to Electromagnetic Radiation', *Med. Pr.,* vol. 47, no. 3, pp. 227–31, 1996.
rf	Nakamura, 1997	In rats exposed to 10 mW/cm^2 the hypothalamic-pituitary-adrenal axis was activated. The results suggest that 'microwaves greatly stress pregnant organisms.'	Nakamura, H. et al., 'Effects of Exposure to Microwaves on Cellular Immunity and Placental Steroids in Pregnant Rats', *Occup. Environ. Med.,* vol. 54, no. 9, pp. 676–80, 1997.
rf	Belousova, 1999	Rats exposed to extremely high frequencies of EMR showed decrease in the weight of adrenal glands.	Belousova, T.E. et al., 'Adrenergic Nerve Plexuses of Heart and Adrenal and Myocardial Catecholamines of Spontaneously Hypertensive Rats Under the Influence of EMR in the Millimeter Range', *Morfologiia,* vol. 115, no. 1, pp. 16–8, 1999.
rf mp	Dasdag, 1999	Rats exposed to radiation from a mobile phone in speech position developed changes to testes.	Dasdag, S. et al., 'Whole-body Microwave Exposure Emitted by Cellular Phones and Testicular Function of Rats', *Urol. Res.,* vol. 27, no. 3, pp. 219–23, Jun. 1999.

Studies relevant to chapter six

EMR reduces production of melatonin/effectiveness of Tamoxifen

vlf	Pfluger, 1996	Railway workers exposed to 16.7 Hz magnetic fields had reduced melatonin levels at night.	Pfluger, D.H. and Minder, C.E., 'Effects of Exposure to 16.7 Hz Magnetic Fields on Urinary 6-hydroxymelatonin Sulfate Excretion of Swiss Railway Workers', *J. Pineal Res.,* vol. 21, no. 2, pp. 91–100, Sep., 1996.
vlf	Karasek, 1998	Men exposed to 40 Hz magnetic fields had significantly reduced levels of melatonin.	Karasek, M. et al., 'Chronic Exposure to 2.9 mT, 40 Hz Magnetic Field Reduces Melatonin Concentrations in Humans', *J. Pineal Res.,* vol. 25, no. 4, pp. 240–4, Dec., 1998.

p	Wilson, 1990	Volunteers exposed to EMR from electric blankets with high fields had reduced levels of melatonin.	Wilson, B.W. et al., 'Evidence for an Effect of ELF Electromagnetic Fields on Human Pineal Gland Function', *J. Pineal Res.*, vol. 9, no. 4, pp. 259–69, 1990.
p	Liburdy, 1993	In breast cancer cells, EMR blocked melatonin's ability to inhibit cancer's growth between 2 and 12 mG.	Liburdy, R.P. et al., 'ELF Magnetic Fields, Breast Cancer, and Melatonin: 60 Hz Fields Block Melatonin's Oncostatic Action on ER+ Breast Cancer Cell Proliferation', *J. Pineal Res.*, vol. 14, no. 2, pp. 89–97, Mar., 1993.
p	Kaune, 1997	Melatonin levels were reduced among people sleeping in higher fields.	Kaune, W. et al., 'Relation Between Residential Magnetic Fields, Light-at-night ad Nocturnal Urine Melatonin Levels in Women', TR-107242-V1, Palo Alto: EPRI, Fred Hutchinson Research Center, 1997.
p	Harland, 1997	EMR counteracts the effects of Tamoxifen in preventing the proliferation of cancer cells.	Harland, J.D. and Liburdy, R. P., 'Environmental Magnetic Fields Inhibit the Antiproliferative Action of Tamoxifen and Melatonin in a Human Breast Cancer Cell Line', *Bioelectromagnetics*, vol. 18, no. 8, pp. 555–62, 1997.
p	Kato, 1997	Rats exposed to a field of 14 mG with circular polarisation had suppressed melatonin; those exposed to 10–50 mG fields with linear polarisation showed no reduction of melatonin.	Kato, M. and Shigemitsu, T. 'Effects of 50Hz Magnetic Fields on Pineal Function in the Rat', in Stevens, R.G. et al (eds), *The Melatonin Hypothesis: Breast Cancer and the Use of Electric Power*, Columbus, Battelle Press, pp. 337–76, 1997.
p	Harland, 1998	EMR counteracts the effects of Tamoxifen in preventing the proliferation of cancer cells.	Harland, J.D. et al., 'Differential Inhibition of Tamoxifen's Oncostatic Functions in a Human Breast Cancer Cell Line by a 12 mG (1.2 uT) Magnetic Field, in *Electricity and Magnetism in*

			Biology and Medicine, Bersani, F. (Ed.), Bologna, Plenum Press, 1998.
p	Wood, 1998	Square-shaped waves produce greater reductions in melatonin than sine waves.	Wood, A.W. et al., 'Changes in Human Plasma Melatonin Profiles in Response to 50 Hz Magnetic Field Exposure', *J. Pineal Res.*, vol. 25, no. 2, pp. 116–27, Sep., 1998.
p	Burch, 1998	Workers in electric utilities with high exposure at home and work had greatest reduction in melatonin levels at night.	Burch, J.B. et al., 'Nocturnal Excretion of a Urinary Melatonin Metabolite Among Electric Utility Workers', Scand. J. Work *Environ. Health*, vol. 24, no. 3, pp. 183–9, Jun., 1998.
p	Anderson and Morris, 1998	Breast cancer cells which had been inhibited by melatonin grew when exposed to a 60 Hz field of 12 mG, but not a field of 2 mG.	Anderson, L. and Morris, J., 20th Annual Meeting of the Bioelectromagnetics Society in St. Pete Beach, Florida, 7–11 Jun. 1998.
p	Burch, 1999	Workers in electric utilities exposed to high levels of EMR had reduced melatonin levels.	Burch, J.B. et al., 'Reduced Excretion of a Melatonin Metabolite in Workers Exposed to 60 Hz Magnetic Fields', *Am. J. Epidemiol.*, vol. 150, no. 1, pp. 27–36, 1 Jul. 1999.
p	Blackman, 2001	EMR counteracts the effects of Tamoxifen in preventing the proliferation of cancer cells.	Blackman, C.F. et al., 'The Influence of 1.2 microT, 60 Hz Magnetic Fields onTamoxifen-induced Inhibition of MCF-7 Cell Growth', *Bioelectromagnetics*, vol. 18, no. 2, pp. 122–8, 2001.
p	Burch, 2000	Men working for more than 2 hours per day in substation or 3-phase environments had reduced levels of melatonin.	Burch, J.B. et al., 'Melatonin Metabolite Levels in Workers Exposed to 60-Hz Magnetic Fields: Work in Substations and with 3-phase Conductors', *J. Occup. Environ. Med.*, vol. 42, no. 2, pp. 136–42, Feb., 2000.

vdu	Afzal, 1996	Magnetic fields from VDUs inhibit oncostatic effect of melatonin.	Afzal, S.M.J. and Liburdy, R. P., 'Magnetic Fields Reduce the Growth Inhibitory Effects of Tamoxifen in a Human Brain Tumor Cell Line' in *Electricity and Magnetism in Biology and Medicine*, Bersani, F. (Ed.), Bologna, Plenum Press, 1998.
rf mp	Burch, 1997	People who reported more frequent use of mobile phones had lower levels of melatonin.	Burch, J.B. et al., 'Cellular Telephone Use and Excretion of a Urinary Melatonin Metabolite', Abstract of the Annual Review of Research on Biological Effects of Electric and Magnetic Fields from the Generation, Delivery & Use of Electricity, San Diego, CA, 1997.

Synergistic effects of EMR and chemicals

p	Loscher, 1993	Rats exposed to chemical DMBA and EMR developed more tumours and larger tumours than those exposed separately.	Loscher, W. et al., 'Tumor Promotion in a Breast Cancer Model by Exposure to a Weak Alternating Magnetic Field', *Cancer Lett.*, vol. 71, nos. 1–3, pp. 75–81, Jul., 1993.
p	Loscher, 1995	Rats exposed long-term to EMR and a chemical carcinogen had an increased incidence of mammary tumours that was related to dose.	Loscher, W. and Mevissen, M., 'Linear Relationship Between Flux Density and Tumor Co-promoting Effect of Prolonged Magnetic Field Exposure in a Breast Cancer Model', *Cancer Lett.*, vol. 96, no. 2, pp. 175–80, 25 Sep. 1995.
p	Mevissen, 1996	Rats treated with carcinogen and EMR developed more mammary tumours than those given EMR or carcinogen separately.	Mevissen, M. et al., 'Exposure of DMBA-treated Female Rats in a 50-Hz, 50 microTesla Magnetic Field: effects on mammary tumor growth, melatonin Levels, and T

			lymphocyte activation', *Carcinogenesis*, vol. 17, no. 5, pp. 903–10, May, 1996.
p	Feychting, 1998	Children exposed to EMR and motor vehicle exhaust had an increased risk of cancers.	Feychting, M. et al., 'Exposure to Motor Vehicle Exhaust and Childhood Cancer', *Scand. J. Work Environ. Health*, vol. 24, no. 1, pp. 8–11, Feb., 1998.
p	Thun-Battersby, 1999	Rats given a carcinogen and exposed to EMR developed more breast cancers than those exposed to stressors separately.	Thun-Battersby, S. et al., 'Exposure of Sprague-Dawley Rats to a 50-Hertz, 100-microTesla Magnetic Field for 27 Weeks Facilitates Mammary Tumorigenesis in the 7,12-dimethylbenz[a]-anthracene Model of Breast Cancer', *Cancer Res.*, vol. 59, no. 15, pp. 3627–33, Aug., 1999.
p	Wachtell, 1999	Children exposed to high traffic and high wire codes had a substantial elevation of childhood cancers.	Wachtell, H. et al., 'Traffic Density and Wire Codes May be Risk Cofactors for Childhood Cancer', *BEMS* 1999, abst 12–3.
rf mpt	Maes, 1996	In blood exposed to EMR from GSM mobile phone towers and carcinogen, there was evidence of chemical mutation.	Maes, A. et al., '954 MHz Microwaves Enhance the Mutagenic Properties of Mitomycin C', *Environ. Mol. Mutagen.*, vol. 28, no. 1, pp. 26–30, 1996.
rf mp	Adey, 1999	Rats exposed to a carcinogen and EMR from a TDMA mobile phone had slightly less tumours than controls.	Adey, W.R. et al., 'Spontaneous and Nitrosourea-induced Primary Tumors of the Central Nervous System in Fischer 344 Rats Chronically Exposed to 836 MHz Modulated Microwaves', *Radiat. Res.*, vol. 152, no. 3, pp. 293–302, Sep., 1999.

Genetic effects

elf	Nordenson	Train drivers exposed to high levels of EMR had twice as many chromosome abnormalities as usual.	Nordenson, I. et al., 'Chromosomal Aberrations in Peripheral Lymphocytes of Train Engine Drivers', BEMS Annual Meeting, Munich, 9–16 June 2000.
p	Lai and Singh, 1997	Rats exposed for 2 hours to a 60 Hz field developed breaks in DNA strands of brain cells.	Lai, H. and Singh, N.P., 'Acute Exposure to a 60 Hz Magnetic Field Increases DNA Strand Breaks in Rat Brain Cells', *Bioelectromagnetics*, vol. 18, no. 2, pp. 156–65, 1997.
p	Svedenstal, 1999	Mice exposed to 50Hz fields from 220 kV line showed DNA damage and changes to leukocytes.	Svedenstal, B.M. et al., 'DNA Damage, Cell Kinetics and ODC Activities Studied in CBA Mice Exposed to Electromagnetic Fields Generated by Transmission Lines', *In Vivo*, vol. 13, no. 6, pp. 507–13, Nov.–Dec., 1999.
p	Chen, 2000	EMF blocked differentiation of red blood cells (blocking is a hallmark of a tumour promotor) which is a genetic effect.	Chen, G. et al., 'Effect of Electromagnetic Field Exposure on Chemically Induced Differentiation of Friend Erythroleukemia Cells', *Environ. Health Perspect.*, vol. 108, no. 10, pp. 967–72, Oct., 2000.
rf	Garaj-Vrhovac, 1990	People occupationally exposed to EMR between 30 and 300 GHz showed greater evidence of chromosome aberrations. (Stewart p. 70)	Garaj-Vrhovac, V. et al., 'Comparison of Chromosome Aberration and Micronucleus Induction in Human Lymphocytes After Occupational Exposure to Vinyl Chloride Monomer and Microwave Radiation', *Periodicum Biologorium*, vol. 92, pp. 411, 1990.

rf mW	Cleary, 1992	Glioma cells exposed to 27 & 2450 MHz showed changes in proliferation and DNA synthesis.	Cleary, S.F. et al., 'Glioma Proliferation Modulated in Vitro by Isothermal Radiofrequency Radiation Exposure', *Radiat. Res.*, vol. 121, no. 1, pp. 38–45, Jan., 1990.
rf	Maes, 1993	Human blood cells exposed to EMR at 2450 MHz showed a marked increase in chromosome aberrations.	Maes, A. et al., 'In Vitro Cytogenetic Effects of 2450 MHz Waves on Human Peripheral Blood Lymphocytes', *Bioelectromagnetics,* vol. 14, no. 6, pp. 495–501, 1993.
rf	Sarkar, 1994	Mice exposed to EMR at 2.45 GHz showed changes in DNA patterns in testis and brain cells.	Sarkar, S. et al., 'Effect of Low Power Microwave on the Mouse Genome: a Direct DNA Analysis', *Mutat. Res.*, vol. 320, nos. 1–2, pp. 141–7, Jan., 1994.
rf	Lai & Singh, 1995	Rats exposed to 2450 MHz developed single-strand DNA breaks in brain cells.	Lai, H. and Singh, N.P., 'Acute Low-intensity Microwave Exposure Increases DNA Single-strand Breaks in Rat Brain Cells', *Bioelectromagnetics*, vol. 16. no. 3, pp. 207–10, 1995.
rf	Lai & Singh, 1996	Rats exposed to 2450 MHz developed single- and double-strand DNA breaks in brain cells.	Lai, H. and Singh, N.P., 'Single- and Double-strand DNA Breaks in Rat Brain Cells After Acute Exposure to Radiofrequency Electromagnetic Radiation', *Int. J. Radiat. Biol.*, vol. 69, no. 4, pp. 513–21, Apr., 1996.
rf	Lai & Singh, 1997	Rats treated with melatonin or PBN had less DNA breaks when exposed to EMR. 'Since both melatonin and PBN are efficient free radical scavengers it is hypothesized that free radicals are involved in RFR-induced DNA damage in the brain cells of rats.'	Lai, H. and Singh, N.P., 'Melatonin and a Spin-trap Compound Block Radiofrequency Electromagnetic Radiation-induced DNA Strand Breaks in Rat Brain Cells', *Bioelectromagnetics*, vol. 18, no. 6, pp. 446–54, 1997.

rf	Lai, 1997	Rats treated with naltrexone before exposure to EMR developed less DNA breaks. This suggests that endogenous opioids help prevent DNA breaks.	Lai, H. et al., 'Naltrexone Blocks RFR-induced DNA Double Strand Breaks in Rat Brain Cells', *Wireless Networks* vol. 3, pp. 471–6, 1997.
rf	Donnellan, 1997	Mast cells exposed to 835 MHz showed changes in growth, morphology and secretion.	Donnellan, M. et al., 'Effects of Exposure to Electromagnetic Radiation at 835 MHz on Growth, Morphology and Secretory Characteristics of a Mast Cell Analogue, RBL-2H3', *Cell Biology International,* vol. 21, no. 7, pp. 427–39, 1997.
rf	Morrissey, 1997	Mice exposed to EMR showed significant increases in the c-fos (cancer-causing) gene.	Morrissey, J.J. et al., 'Effects of 1.6 GHz Microwaves (CW and Pulsed Wave) on c-fos EGFR and NSCL-1 Gene Expression in the Mouse Brain', *Proceedings of the 1997 World Congress on Electricity and Magnetism in Biology and Medicine,* 1997.
rf	Phillips, 1998	Cells exposed to EMR showed increases and decreases in DNA damage.	Phillips, J.L. et al., 'DNA Damage in Molt-4 T-lymphoblastoid Cells Exposed to Cellular Telephone RFs in Vitro', *Bioelectrochemistry and Bioenergetics,* vol. 45, pp. 103–10, 1998.
rf mp	Daniells, 1998	Worms susceptible to cancer showed a stress response to EMR at 700 MHz and 300 MHz, with lower power giving greater response.	Daniells, C. et al., 'Transgenic Nematodes as Biomonitors of Microwave-induced Stress', *Mutat. Res.,* vol. 399 no. 1, pp. 55–64, 13 Mar. 1998.
rf	Goswami, 1999	Cells exposed to EMR from mobile phones showed changes to the c-fos gene which is involved in cell proliferation and differentiation associated with cancer.	Goswami, P.C. et al., 'Expression of Proto-oncogene and Activities of Multiple Transcription Factors in RF Exposed Cells, Using C3H10T1/2 Mouse Embryo Fibroblast Cells Exposed to 835.62 and 847.74 MHz Cellphone Radiations', *Radiat. Res.*. vol. 151, no. 3, pp. 300–9, 1999.

rf	Goswami, 1999	Cells exposed to EMR at 2 mobile frequencies expressed a higher level of genes (proto-oncogenes) which can produce cancer.	Goswami, P.C. et al., 'Changes in Embryonic Fibroblasts Exposed to Cellphone Radiations', *Radiation Research*, vol. 151, no. 3, pp. 300–9, Mar., 1999.
rf	Sarkar, 1999	In mice exposed to 2450 MHz there was a rearrangement of DNA in the cells of testis and brains.	Sarkar, S., 'Tandem Repeat Sequences as Markers to Study Microwave-DNA Interaction', National Seminar on Low-level Electromagnetic Field Phenomena in Biological Systems, New Delhi, India, Feb., 1999.
rf mp	Hook, 1999	Cells exposed to EMR from mobile phones showed evidence of DNA disruption.	Hook, G.J. et al., 'Genotoxicity of RF Fields Generated From Analog, TDMA, CDMA and PCS Technology', *BEMS*, Long Beach, California, 1999.
rf mp	Harvey, 2000	In mast cells exposed to 864.3 MHz, three genes were affected.	Harvey, C. and French, P.W., 'Effects on Protein Kinase C and Gene Expression in a Human Mast Cell Line, HMC-1, Following Microwave Exposure', *Cell Biol. Int.*, vol. 23, no. 11, pp. 739–48, 2000.

Appendix B

Council policies
for reducing exposure

Moreland City Council Strategy for Reducing Public Exposure to Electromagnetic Fields

September 1998, Draft

While scientific opinion is divided on the degree of certainty and on the precise mechanisms by which EMFs interact with living cells in such a way as to cause health problems there is a consensus that it is prudent to avoid and reduce exposure.

Moreland notes the recommendations of the NSW Gibbs Inquiry, the Victorian Powerline Review Panel and recent research findings, and this strategy recommends adopting 'prudent avoidance' as a guiding philosophy in all electricity and radio-communications infrastructure planning.

It is further recommended that Council work towards the achievement of the following standards:

- long-term human habitat is subject to no more than 2 mG ambient extremely low frequency EMFs. This is a long term target of the lowest feasible field strength.
- average ambient extremely low frequency EMFs in the City of Moreland work locations should not exceed 2 mG, as a long-term target.
- radio frequency radiation ambient power flux density in general public areas of the city, and world locations, does not exceed an average of 2 microwatts per square centimetre (μW/cm2).

The 2 mG limit has been chosen as it is the lowest practical field strength achievable—it is also the lowest level at which health effects have been associated with EMFs.

The electricity industry official policy is prudent avoidance so the immediate aim is to achieve agreement with the power companies on elements of a prudent avoidance strategy in Moreland and divide the City's strategy into immediate and long-term objectives.

Moreland City Council recognises that this is a 20-year project and will stay actively engaged with EMFs and related planning issues to ensure that programs are embedded in recurrent funding. ...

As it is Council's intention to minimise the risk of long-term exposure, a planning philosophy of prudent avoidance is recommended for all development and planning purposes.

The prudent avoidance approach should be applied in those residential situations where residents, in particular children, could be subject to long-term exposure.' (pp. 2–3.)

[The substance of many of the recommendations in this document have been included in this book.]

Sutherland Shire Council Local Guideline for Acceptability of Siting Microwave Base Stations
January 1997

*Zone/Use**

Industrial/business	300 metres from any residence unless annual average exposure at nearest residence is less than $0.20 +/- 0.02 \mu W/cm^2$.
Dwellings/boarding houses/residential land/sensitive uses	300 metres from any dwelling, boarding house or residential land unless annual average exposure at any of these is less than $0.20 +/- 0.02 mW/cm^2$
	300 metres from any school, childcare centre, hospital or aged care centre and less than $0.20 +/- 0.02 \mu W/cm^2$ annual average exposure.
Bushland/wetland/ national park/nature reserves and state recreation areas	only with consideration of environmental procedures to minimise impact during construction, maintenance and operation, and also with an acceptable workplan.

* Industrial zones are the preferred location for base stations

Abbreviations

AC	alternating current
ACA	Australian Communications Authority
ACTU	Australian Council of Trade Unions
ARPANSA	Australian Radiation Protection and Nuclear Safety Agency
CDMA	Code division multiple access (a digital mobile phone technology)
CSIRO	Commonwealth Scientific & Industrial Research Organisation
DC	Direct current
ELF	Extra low frequency
EMF	Electromagnetic fields
EMR	Electromagnetic radiation
EMRAA	Electromagnetic Radiation Association of Australia (peak community group)
EPA	Environmental Protection Authority
ESAA	Electricity Supply Authority of Australia (peak industry group)
GSM	Global System for Mobile Communications; a pulsed digital mobile phone system
ICNIRP	International Commission on Non-Ionizing Radiation Protection
MW News	*Microwave News*
$\mu W/cm^2$	Microwatts per square centimetre
NIEHS	National Institute of Environmental Health Sciences (US)
RAPID	Electric and Magnetic Fields Research and Public Information Dissemination Program, 1992–9 (US)
RF	radiofrequency
SAR	specific absorption rate
SQUID	Super-conducting quantum interference device
TDMA	Time Division Multiple Access, a mobile phone system which allows each channel to be used by several mobile phones
UV	ultra violet
VDU	Video Display Unit
WHO	World Health Organisation

Resources

EMR Association of Australia

The EMR Association provides a quarterly newsletter on the health effects of EMR.
PO Box 589
Sutherland 1499
(02) 9523 4750 and (02) 9545 3077
Email: emraa@acay.com.au
web page: http://ssec.org.au/emraa

EMR Surveys

The EMR Association has endorsed the services of a number of qualified electrical engineers who conduct thorough, independent EMR surveys:

New South Wales
Details of consultants working in different areas can be obtained from the Association's office at 02 9523 4750

Victoria
Brunswick Energy and Environmental Services Pty Ltd
Roger Lamb
11 Lavender Park Rd,
Eltham 3095
ph 0419 582 018; fax (03) 9431 1811
email: Lamze@netspace.com.au

Computers—field-reducing screens

Raptek
Unit 12

171 Gibbes St
Chatswood NSW 2067
Ph: 1300 727 835
raptek@raptek.com.au

Fluorescent light reflectors

Silverlux reflectors amplify light from fluorescent fittings
Environment and Safety Service
238 Parramatta Rd, Stanmore 2048
phone: 02 9560 9789
fax: 02 9560 9799
email: essretrofitters@ozemail.com.au
www.ozemail.com.au/~essretrofitters

Medical

Dr Bruce Hocking specialises in occupational and environmental medicine
Suite 401, 34 Queens Rd, Melbourne
03 9866 6376

Power cables

Twisted, shielded VMVB power cables are available from
Pirenne and Ooms
Z1 des Plenesses
rue de l'Avenir 5
Thimister Clermont
Belgium
Tel: 32 87 445 345; fax: 32 87 445 822

Shielding specialists

RFI Industries Pty Ltd
Specialists in shielding from power and radiofrequency sources
54 Holloway Drive
Bayswater Vic 3153
Tel: 03 9762 6733; fax: 03 9762 2247
Email: sales@rfi-ind.com.au
Web page: rfi-ind.com.au

Magshield
Specialises in mitigation and shielding of 50 Hz EMF emitted by sources such as power equipment, substations, power lines, and cables.
Garry Melik
Magshield
6/1 Edinburgh St
Hampton Vic 3188
Tel: 03 9521 6068
Fax: 03 9598 0328
Email: gmelik@magshield.com.au
Web page: www.magshield.com.au

Glossary

Analogue
This is a continuous electromagnetic wave which reproduces an audio signal. This technique was incorporated into the original mobile phone technology.

Atom
An atom consists of a positively charged protons, negatively charged electrons and neutrons with 0 charge.

Base station
This is a mobile phone tower or antenna with associated infrastructure.

Carrier wave
This is a wave which carries superimposed information.

Chromosome
These are components of a cell's nucleus which contain genes.

Circuit
This is the route electricity takes from the point of supply to the load (for example, an appliance) and back to the source.

Colocation
The location of an additional base station or antenna on an existing base station by another carrier.

Conductor
This is a material, such as metal, that allows the easy flow of electrons.

Control
A control is similar to the person, cell or animal under study. However, because it does not have the disease or it is not exposed to the experimental conditions, it functions as a comparison.

Current
Current is the flow of electrons around a circuit (such as the power grid). It is measured in amps and creates an electromagnetic field.

Alternating current (AC)
This is current that regularly reverses its direction. 50-cycle (50 Hz) AC is an electrical current that changes its direction of flow 50 times per second. Graphed, alternating current forms a sine wave like this:

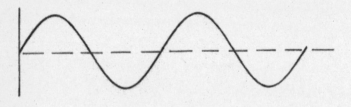

Digital
This is a series of pulses which reproduce an audio signal. This technique is now generally used in mobile phone technologies.

Direct current (DC)
DC is a steady electrical current which does not change direction like alternating current.

Earthing
This is the tendency of electricity to flow—via the easiest route—to earth.

Electric field
This field is the result of voltage and is present whenever an appliance is turned on, even if it is switched off. It is measured in volts per metre or variations such as microvolts per centimetre.

Electromagnetic field (EMF)/ Electromagnetic radiation (EMR)
EMF/R is an energy that radiates from any moving electrical current. It contains both a magnetic field and an electrical field.

Electromagnetic spectrum (EM spectrum)
This is the organisation of radiation according to its frequency. The non-ionising electromagnetic spectrum contains DC, ELF, RF and microwaves up to visible light. The ionising part of the spectrum includes x-rays, cosmic rays and nuclear radiation.

Electrons
These are one of the three components of an atom, the others being protons (positively charged) and neutrons (neutral). Electrons have a negative charge.

Epidemiology
These are studies of human populations exposed to a particular agent and include research on people living near high voltage powerlines or working in particular occupations.

Extra Low Frequency (ELF)
ELF is the portion of the electromagnetic spectrum from zero to 1,000 Hz. This includes the 50-cycle power used in Australia.

Frequency
This meansures the number of wavelengths to pass a given point in a given period. It is normally measured in Hertz, which is the number of wavelengths passing a given point in one second. Higher frequencies have shorter wavelengths than lower frequencies.

Gauss (G)
This is a unit of intensity of a magnetic field. 1G = 1,000 mG (milliGauss)

Genes
These units contain our hereditary characteristics and are attached to chromosomes.

Geomagnetic field
The earth exerts an electromagnetic field of 0.5 Gauss.

Grounding

Electricity returns to earth via the easiest route, be it water pipes, the return wire, earth stake or an unfortunate person. Insulators are placed on the larger distribution towers to prevent current flowing to earth via the tower itself.

Hertz

This is a measure of frequency (the number of cycles per second), named after Heinrich Hertz, German physicist who discovered radio waves in 1888.
KiloHertz (KHz) = a thousand Hertz;
Megahertz (MHz) = a million Hertz;
GigaHertz (GHz) = a thousand million Hz
In Australia, electricity operates at 50 Hz (50 cycles per second).

In vitro

Latin for 'in glass', this phrase refers to laboratory studies that are conducted in a test tube outside the body of an organism.

In vivo

Latin for 'in a living thing', this phrase refers to laboratory studies conducted on animals, usually rats or mice.

Ionising radiation

Electromagnetic radiation with the energy to break chemical bond is known as ionising radiation. X-rays and nuclear radiation come within this category.

Ion

An ion is an atom that has lost or gained one or more electrons. An atom that has lost an electron/electrons becomes a positively charged ion; an atom that has gained an electron/electrons from elsewhere becomes a negatively charged ion.

Magnetic field

A magnetic field exists only when current is flowing. It is not blocked by solid objects, and travels through nearly all materials. It is measured in milliGauss (mG) in Europe and Australia, or microTesla (uT) in the US. 1 uT = 10mG.

Melatonin

This hormone, produced by the pineal gland, is responsible for many functions (see chapter 6), including acting as a free radical scavenger.

Microwatts/cm² (μW/cm²)
This measures. the power density of mobile phones and base stations.

Microwave
This is the portion of the electromagnetic spectrum from 500 MHz up to the frequency of visible light. Technology that operates at this frequency includes radar, telecommunications links, satellite communications, weather-observation equipment, and medical diathermy.

Modulated wave
A modulation is a change in frequency or strength of the field. A pulsed wave is a modulated wave which has bursts of energy. A modulated wave carries information and is superimposed on a carrier wave.

Amplitude modulation
This is a wave that has been modulated (changed) in amplitude (size/shape) but not frequency.

Frequency modulation
These are waves that are modulated (changed) in frequency, but their amplitude (size or shape) but remains the same.

Non ionising radiation
This is that part of the electromagnetic spectrum extending from zero frequency to frequencies of visible light. This radiation does not contain sufficient energy to break chemical bonds.

Polarisation
This refers to the orientation of a field through space. A linearly polarised field may be oriented vertically or horizontally, or a combination of the two. A circularly polarised field occurs when the field is propagated as a series of circles. (See the hose analogy on p. 91.)

Radio Frequency (RF)
RF is that portion of the electromagnetic spectrum from 500 kHz to 500 MHz.

Resistance
This is the force that restricts (or opposes) the flow of electricity such as a non-

conductive material or a thin wire.

Resonance
Resonance occurs when the length of a wave is the same size as an object and imparts its energy to the object. A well-known example is the opera singer who reaches a pitch at which glass shatters. A standing adult resonates to about 77 MHz.

Semiconductors
Half way between conductors (which allow easy flow of electrons) and insulators (which impede their flow), semiconductors allow small currents to be transported over long distances and to be controlled precisely.

Specific Absorption Rate (SAR)
This is the rate at which microwaves are absorbed by a mass, usually the body. The SAR will be affected by the size and density of the body and the frequency of the field. It is measured in watts/milliwatts or microwatts per square metre.

Static electricity
In the electricity that powers our homes, electrons move through a conductor. In static electricity, however, the charges remain on the surface of the object without moving.

Transformers
A transformer converts electricity to a different voltage. It achieves this conversion by using coils of wire within which the appropriate number of twists induces a different voltage and current.

Very Low Frequency (VLF)
VLF is that portion of the electromagnetic spectrum from 1kHz to 500 kHz.

Voltage
Voltage is the electrical force which 'pushes' electrons through the circuit. It is measured in volts and causes an electric field which is measured in volts per metre (V/m).

Wavelength
This is the distance from one point on a wave to the identical point on the next wave. Higher frequencies have shorter wavelengths than lower frequencies.

References

ACTU OHS Unit, *ACTU Guidelines for Screen Based Work,* 393 Swanston St, Melbourne, 3000, May 1998.

Adey, W.R., correspondence to Don Maisch, Aug. 1995.

——. in Kuster, N., Balzano, Q. and Lin, J.C. (eds.), *Progress in Safety Assessments of Mobile Communications*, Chapman Hall, New York, 1996.

Alfredsson, L. et al., 'Cancer Incidence Among Male Railway Engine-drivers and Conductors in Sweden, 1976-90', *Cancer Causes Control*, vol. 7, no. 3, pp. 377–81, 1996.

Amtsgericht München, Aktenzeichen 432 C 7381/95, 27 Mar. 1998.

Ashton, J. and Laura, R., *Perils of Progress*, UNSW Press, Sydney, 1997.

Australian Radiation Laboratory, Information Sheets, 1993.

Barbaro, V. et al., 'Electromagnetic Interference of Analog Cellular Telephones with Pacemakers', *Pacing and Clinical Electrophysiology* vol. 19, no. 10, pp. 1410–18, Oct. 1996.

Baris, D. et al., 'A Case Cohort Study of Suicide in Relation to Exposure to Electric and Magnetic Fields Among Electricity Utility Workers', *Occup. Environ. Med.*, vol. 53, no. 1, pp. 17–24, 1996.

Barnett, S., *Status of Research on Biological Effects and Safety of Electromagnetic Radiation: telecommunications frequencies*, CSIRO, 1994.

Bawin, S.M. and Adey, W.R., 'Sensitivity of Calcium Binding in Cerebral Tissue to Weak Environmental Electric Fields Oscillating at Low Frequency', *Proc. Natl Acad. Sci. USA,* vol. 73, no. 6, pp. 1999–2003, 1976.

Bawin, S.M. et al., 'Effects of Modulated VHF Fields on the Central Nervous System', *Ann. NY Acad. Sci.*, vol. 274, p. 74, 1975.

Beale, I.L., 'Psychological Effects of Chronic Exposure to 50-Hz Magnetic Fields in Humans Living Near Extra-High-Voltage Transmission Lines', *Bioelectromagnetics,* vol. 18, no. 8, pp. 584–94, 1997.

Becker, R. O. and Seldon, G., *The Body Electric*, William Morrow, New York, 1985.

Becker, Robert O., *Cross Currents: The Perils of Electropollution, The Promise of Electromedicine*, Tarcher/Putnam, New York, 1990.

Belanger, K. et al., 'Spontaneous Abortion and Exposure to Electric Blankets and Heated Water Beds', *Epidemiology,* vol. 9, no. 1, pp. 36–42, Jan. 1998.

Bell, G.B. et al., 'Frequency-specific Responses in the Human Brain Caused by Electromagnetic Fields', *Journal Neurological Science,* vol. 123 nos 1–2, pp. 26-32, May 1994.

Beral, V. et al., 'Malignant Melanoma and Exposure to Fluorescent Lighting at Work', *Lancet,* vol. 2, no. 8293, pp. 290-93, 7 Aug. 1982.

Blackman, C.F. et al., 'Multiple Power-density Windows and their Possible Origin', *Bioelectromagnetics,* vol. 10, no. 2, pp. 115–28, 1989.

Blank, M. and Goodman, R., 'Do Electromagnetic Fields Interact Directly with DNA?' *Bioelectromagnetics,* vol. 18, no. 2, pp. 111–15, 1997.

Bonhomme-Faivre, L. et al., 'Study of Human Neurovegetative and Hematologic Effects of Environmental Low-Frequency (50 Hz) Electromagnetic Fields Produced by Transformers', *Archives of Environmental Health,* vol. 53, no. 2, pp. 87–92, Mar.-Apr. 1998.

Brix, J et al., 'Analysis of Individual Exposure Data of a Representative Group of 2000 People to Low Frequency Magnetic Fields Over 24 H', abstract 8-4 at BEMS conference, 1999.

Brodeur, P., *Currents of Death,* Simon and Schuster, New York, 1989.

Butler, J., Didriksen, N. and Harrell, E.H., 'Neurocognitive Patterns of Chemical Sensitivity', 1977 Annual International Symposium on Man and His Environment in Health and Disease.

Cellular Phone Taskforce, *No Place to Hide,* newsletter, PO Box 100404, Brooklyn, New York 11210.

Cherry, N., *Potential and Actual Adverse Effects of Radiofrequency and Microwave Radiation at Levels Near and Below 2 Microwatts/sq. cm.*, Lincoln University New Zealand, 1997.

——. *Criticism of the Health Assessment in the ICNIRP Guidelines for Radiofrequency and Microwave Radiation (100kHz–300 GHz)*, Lincoln University, NZ, 22 Oct. 1999. [available at www.nzine.co.nz/thesis.html]

Cherry, N., Interview by Linda Howe, on Art Bell Dreamland Show, US, 30 Jan. 2000.

Coleman, M.P. et al., 'Leukaemia and Residence Near Electricity Transmission Equipment: a Case-control Study', *Br. J. Cancer*, vol. 60, no. 5, pp. 793–8, Nov., 1989.

Colwell, P.F., 'Powerlines and Land Value', *J. of Real Est. Rsch.*, vol. 5, no. 1, pp. 117–127, Spring 1990.

Coogan, P.F. et al., 'Occupational Exposure to 60-Hertz Magnetic Fields and Risk of Breast Cancer in Women', *Epidemiology,* vol. 7, no. 5, pp. 459–64, Sept. 1996.

Cowan, D. and Girdlestone, R., *Safe As Houses,* Gateway, UK, 1995.

Dalton, L., *Radiation Exposures*, Scribe, Victoria, 1991

Daniells, C. et al., 'Transgenic Nematodes as Biomonitors of Microwave-induced Stress', *Mutat-Res.*, vol. 399 no. 1, pp. 55–64, 13 Mar. 1998.

Davanipour, Z. et al., 'Amyotrophic Lateral Sclerosis and Occupational Exposure to Electromagnetic Fields', *Bioelectromagnetics,* vol. 18, no. 1, pp. 28–35, 1997.

Delgado, J.M.R., 'Physical Control of the Mind: Toward a Psychocivilized Society', *World Perspectives*, vol. 41, Harper and Row, New York, 1969.

Demers, P.A. et al., 'Occupational Exposure to Electromagnetic Fields and Breast Cancer in Men', *Am. J. Epidemiol.*, vol. 134, no. 4, pp. 340–7, 15 Aug. 1991.

de Pomerai, D. et al., 'Non-thermal Heat-shock Response to Microwaves, *Nature*, vol. 405, 25 May 2000.

Dolan, M., Nuttall, K., Flanagan, P. And Melik, G., 'The Application of Prudent Avoidance in EMF Risk Management', presented to WHO International Seminar on EMF Risk Perception and Communication, Ottawa, 31 Aug.–4 Sept. 1998.

Donnellan, M. et al., 'Effects of Exposure to Electromagnetic Radiation at 835 MHz on Growth, Morphology and Secretory Characteristics of a Mast Cell Analogue, RBL-2H3', *Cell Biol Int.*, vol. 21, pp. 427–39, 1997.

Doull, A.H. and Curtain, C., *Radiofrequency Radiation Exposure Standards in Australia and New Zealand—A Case for Reducing Human Exposure Limits Based on Low Level, Non Thermal, Biological Effects*, CSIRO, 1994.

Electrical Sensitivity News, formerly a bi-monthly newsletter from Weldon Publishing, PO Box 4146, Prescott AZ, 86302 USA.

Electromagnetic Hazard and Therapy, 1997/8, vol. 8, nos. 3–4.

EMRAA, *EMRAA News,* a quarterly newsletter on electromagnetic radiation, PO Box 589, Sutherland 1499, Australia

ESAA Information Sheets 1–6, 1994.

Fews, A.P., Henshaw, D.L.,et al., 'Increased Exposure to Pollutant Aerosols Under High Voltage Power Lines', *Int. J. Radiat. Biol.*, vol. 75, no. 12, pp 1505–21.

Fews, A.P., Henshaw, D.L. et al., 'Corona Ions from Powerlines and Increased Exposure to Pollutant Aerosols', *Internat. J. Radiat. Biol,.* vol. 75, no. 12, p. 1523–31.

Feychting, M. et al., 'Occupational and Residential Magnetic Field Exposure and Leukemia and Central Nervous System Tumors*', Epidemiology,* vol. 8, no. 4, pp. 384–389, Jul. 1997.

——.'Exposure to Motor Vehicle Exhaust and Childhood Cancer', *Scand. Journal of Work and Environ. Health,* vol. 24, no. 1, pp. 8–11, Feb. 1998.

——.'Dementia and Occupational Exposure to Magnetic Fields', *Scand. Journal Work and Environ. Health,* vol. 24, no. 1, pp. 46–53, Feb. 1998.

Feychting, M. and Ahlbom, A., 'Magnetic Fields and Cancer in Children Residing Near Swedish High-Voltage Power Lines', *Am. J. Epidemiol.,* vol. 138, no. 7, pp. 467–81, 1 Oct. 1993.

Finkelstein, M.M., 'Cancer Incidence Among Ontario Police Officers', *Am. J. Ind. Med.,* vol. 34, no. 2, pp. 157–62, Aug. 1998.

First World Conference on Breast Cancer, Canada, July 1997.

Firstenberg, A., *Microwaving Our Planet*, self-published, New York, 1996.

Floderus, B., et al., 'Occupational Exposure to Electromagnetic Fields in Relation to Leukemia and Brain Tumors: a Case-control Study in Sweden', *Cancer Causes Control,* vol. 4, no. 5, pp. 465–76., Sept. 1993.

——. 'Incidence of Selected Cancers in Swedish Railway Workers, 1961-79', *Cancer Causes Control,* vol. 5, no. 2, pp. 189-94, Mar. 1994.

——. 'Occupational Magnetic Field Exposure and Site-specific Cancer Incidence: a Swedish Cohort Study', *Cancer Causes Control,* vol. 10, no. 5, pp. 323–32, Oct. 1999.

Florig, Keith (Dr) and Richard Barry, *VDT News*, Jan./Feb. 1992, p. 2.

Forskning and Praktik, 'WWDU Problems Have Many Causes', Swedish National Insdtitute of Occupational Health, vol. 4, 1992.

French, P.W., *Biological Effects of Environmental Electromagnetic Fields from Mobile Phone Towers—A Summary of What We Know and of What We Need to Know*, Centre for Immunology, St Vincent's Hospital, Sydney, NSW 2010, Australia, Jun. 1997.

French, P.W. et al., 'Electromagnetic Radiation at 835 MHz Changes the Morphology and Inhibits Proliferation of a Human Astrocytoma Cell Line', *Bioelectrochemistry and Bioenergetics,* vol. 43, pp. 13–18, 1997.

Frey, A.H., 'Electromagnetic Field Interactions with Biological Systems', *FASEB Journal,* vol. 7, no. 2, pp. 272–81, Feb. 1993.

——. 'Headaches From Cellular Telephones: Are They Real and What Are the Implications?', *Environ. Health Perspect.*, vol. 106, no. 3, pp. 101–3, 1998.

Gandhi, O.P. et al., 'Electromagnetic Absorption in the Human Head and Neck for Mobile Telephones at 835 and 1900 mHz', *IEEE Transactions on Microwave Theory and Techniques*, vol. 44, no. 10, Oct. 1996.

Gangi, S. and Johansson, O., 'A Theoretical Model Based Upon Mast Cells and Histamine to Explain the Recently Proclaimed Sensitivity to Electric and/or Magnetic Fields in Humans', *Medical Hypotheses*, vol. 54, no. 4, pp. 663–71, 2000.

Gibbs, Harry, *Inquiry into Community Needs and High Voltage Transmission Line Development*, 1990, New South Wales Government Printing Office.

Goldhaber, M.K. et al., 'The Risk of Miscarriage and Birth Defects Among Women Who Use Visual Display Terminals During Pregnancy', *Am. J. Ind. Med.,* vol. 13, no. 6, pp. 695–706, 1988.

Goldsmith, J., 'Epidemiologic Evidence of Radiofrequency Radiation (Microwave) Effects on Health in Military, Broadcasting, and Occupational Studies', *International J. of Occup. and Environmental Health*, vol. 1, no. 1, pp. 47–57, 1995.

Grant, L., *The Electrical Sensitivity Handbook—How Electromagnetic Fields (EMFs) Are Making People Sick*, Weldon Publishing, 1995.

——. *Workstation Radiation*, Weldon Publishing, Arizona, 1992.

Green, L.M. et al., 'A Case-control Study of Childhood Leukemia in Southern Ontario, Canada, and Exposure to Magnetic Fields in Residences', *Int. J. Cancer*, vol. 82, no. 2, pp.161–70, Jul. 1999.

Gregory, R and von Winterfeldt, D., 'The Effects of Electromagnetic Fields from Transmission Lines on Public Fears and Property Values', *J. of Environmental Mgt.*, vol. 48 pp. 201–14, 1996.

Guenel, P. et al., 'Exposure to 50-Hz Electric Field and Incidence of Leukemia, Brain Tumors, and Other Cancers Among French Electric Utility Workers', *Am. J. Epidemiol.,* vol. 144, no. 12, pp. 1107–21, 15 Dec. 1996.

Ham, W.T., 'Ocular Hazards of Light Sources: Review of Current Knowledge', *J. Occup. Med.*, vol. 25, no. 2, pp. 101–3, Feb. 1983.

Hardell, L. et al., 'Angiosarcoma of the Scalp and Use of a Cordless (Portable) Telephone', *Epidemiology,* vol.10, no. 6, pp. 785–6, 1999.

Harvey, C. and French, P.W., 'Effects on Protein Kinase C and Gene Expression in a Human Mast Cell Line, HMC-1, Following Microwave Exposure', *Cell Biol. Int.,* vol. 23, no. 11, pp. 739–48, 2000.

Hatch, E.E. et al., 'Association Between Childhood Acute Lymphoblastic Leukemia and Use of Electrical Appliances During Pregnancy and Childhood', *Epidemiology,* vol. 9, no. 3, pp. 234–45, May 1998.

Hayes, D.L. et al., 'Interference with Cardiac Pacemakers by Cellular Telephones', *N. Eng. J. Med.*, vol. 336, no. 21, pp. 1473–79, 22 May 1997.

Hertel, H. and Blanc, B., *Hidden Hazards of Microwave Cooking,* Acres, USA, Apr. 1994.

Hutchison, R., Ph.D., 'Economic Impact Study, Property Value Declines Associated with the Perceived Medical Harm from a Proposed High Definition Television Broadcast Antenna', Colorado, 3 Mar. 1999.

Hyland, G J, 'Physics and Biology of Mobile Telephony', *Lancet*, vol. 356, pp. 1833–6, 2000.

Hyman, J.W., *The Light Book—How Natural and Artificial Light Affect Our Health, Mood, and Behaviour*, Tarcher, Los Angeles, 1990.

Independent Expert Group on Mobile Phones, Sir William Stewart (Chairman), *Mobile Phones and Health*, c/- National Radiological Protection Board, Chilton, UK, 2000. The full text of the report can be found at http://www.iegmp.org/uk.

Infante-Rivard, C., 'Electromagnetic Field Exposure During Pregnancy and Childhood Leukemia', *Lancet,* vol. 346, no. 8968, pp. 177, 15 July 1995.

Inquiry into Electromagnetic Radiation, Report of the Senate Environment, Communications, Information Technology and the Arts References

Committee, Parliament House, Canberra, May 2001.

Inskip, P.D. et al., 'Cellular-Telephone Use and Brain Tumors', *N. Engl J. Med.*, vol 344, no. 2, pp. 79-86, 2001.

Irnich, W. and Tobisch, R., 'Mobile Phones in Hospitals', *Biomedical Instrumentation and Technology*, vol. 33, no. 1, pp. 28–43, Jan.–Feb. 1999.

Ji, B.T. et al., 'Occupation and Pancreatic Cancer Risk in Shanghai, China', *Am. J. Ind. Med.,* vol. 35, no. 1, pp. 76–81, Jan. 1999.

Johansen, C. and Olsen, J., 'Mortality from Amyotrophic Lateral Sclerosis, Other Chronic Disorders, and Electric Shocks Among Utility Workers', *Am. J. Epidemiol.*, vol. 148, no. 4, pp. 362–8, 15 Aug. 1998.

Johansen, C. et al., 'Cellular Telephones and Cancer — a Nationwide Cohort Study in Denmark', *Journal of the National Cancer Institute*, vol. 93, pp. 203–7, 2001.

Johansson, O. et al., 'Skin Changes in Patients Claiming to Suffer From 'Screen Dermatitis': a Two-case Open-field Provocation Study', *Exp. Dermatol.*, vol. 3, no. 5, pp. 234–38, Oct. 1994.

——. 'A Screening of Skin Changes, With Special Emphasis on Neurochemical Marker Antibody Evaluation, in Patients Claiming to Suffer From 'Screen Dermatitis' as Compared to Normal Healthy Controls', *Exp. Dermatol.*, vol. 5, no. 5, pp. 279–85, 1996.

Joines, W.T. et al., 'Microwave Absorption Power Differences Between Normal and Malignant Tissue', *Int. J. Rad. Onc., Biol. and Phys.*, vol. 6, pp. 681–7, 1980.

Juutilainen, J. et al., 'Incidence of Leukaemia and Brain Tumours in Finnish Workers Exposed to ELF Magnetic Fields', *Int. Arch. Occup. Environ. Health,* vol. 62, no. 4, pp. 289–93, 1990.

Kayrooz, C., Kinnear, P. and Preston, P., *Academic Freedom and Commercialisation of Australian Universities: Perceptions and Experiences of Social Scientists*, Discussion Paper Number 37, Mar. 2001.

Kirschvink, J.L., 'Magnetite in Human Tissues: a Mechanism for the Biological Effects of Weak ELF Magnetic Fields', *Bioelectromagnetics*, Suppl. 1, pp. 101–13, 1992.

——.'Microwave Absorption by Magnetite: a Possible Mechanism for Coupling Nonthermal Levels of Radiation to Biological Systems', *Bioelectromagnetics,* vol. 17, no. 3, pp. 187–94, 1996.

Kwee, S. and Rasmark, P., 'Radiofrequency Electromagnetic Fields and Cell Proliferation', Second World Congress for Electricity and Magnetism in Biology and Medicine, Bologne, June 1997.

Lai, H., 'Neurological Effects of Radiofrequency Electromagnetic Radiation Relating to Wireless Communication Technology', paper presented at IBC-UK Conference, 'Mobile Phones—Is there a Health Risk', Brussels, Belgium, 16–17 Sept. 1997.

——. 'Neurological Effects of Radiofrequency Electromagnetic Radiation, paper presented to the Workshop on Possible Biological and Health Effects of RF Electromagnetic Fields', Vienna, Austria, 25–28 Oct. 1998.

Lai, H. and Singh, N.P., 'Acute Low-Intensity Microwave Exposure Increases DNA Single-strand Breaks in Rat Brain Cells', *Bioelectromagnetics,* vol. 16 no. 3, pp. 207–10, 1995.

——. 'Single- and Double-strand DNA Breaks in Rat Brain Cells After Acute Exposure to Radiofrequency Electromagnetic Radiation', *Int. J. Radiat. Biol.*, vol. 69, no. 4, pp. 513–21, Apr. 1996.

——. 'Acute Exposure to a 60 Hz Magnetic Field Increases DNA Strand Breaks in Rat Brain Cells', *Bioelectromagnetics,* vol. 18, no. 2, pp. 156–65, 1997.

Laurence, J.A. et al., 'Biological Effects of Electromagnetic Fields—Mechanism for the Effects of Pulsed Microwave Radiation on Protein Conformation', *J. Theor. Biol.*, vol. 206, no. 2, pp. 291–8, Sept. 2000.

Leel, G.M. et al., 'Preliminary Description of 6-Hydroxymelatonin Sulfate Levels by Menstrual Cycle Phase, Day Length, Electric Blanket Use', Abstract 2-4, 1999 BEMS Meeting.

Li, C.Y. et al., 'Residential Exposure to 60-Hertz Magnetic Fields and Adult Cancers in Taiwan', *Epidemiology,* vol. 8, no. 1, pp. 25–30, Jan. 1997.

——. 'Risk of Leukemia in Children Living Near High-voltage Transmission Lines', *J. Occup. Environ. Med.*, vol. 40, no. 2, pp. 144–7, Feb. 1998.

Liburdy, R.P. et al., 'ELF Magnetic Fields, Breast Cancer, and Melatonin: 60 Hz Fields Block Melatonin's Oncostatic Action on ER+ Breast Cancer Cell Proliferation', *Journal of Pineal Research,* vol. 14, no. 2, pp. 89–97, Mar. 1993.

Lin, R.S. et al., 'Occupational Exposure to Electromagnetic Fields and the Occurrence of Brain Tumors. An Analysis of Possible Associations', *J. Occup. Med.*, vol. 27, no. 6, pp. 413–19, June 1985.

Linet, M.S. et al., 'Residential Exposure to Magnetic Fields and Acute Lymphoblastic Leukemia in Children', *N. Eng. J. Med.,* vol. 337, no. 1, pp. 1–7, 3 July 1997.

London, S.J. et al., 'Exposure to Residential Electric and Magnetic Fields and Risk of Childhood Leukemia', *Am. J. Epidemiol.,* vol. 134, no. 9, pp. .923–37, 1 Nov. 1991.

Loomis, D.P. et al., 'Breast Cancer Mortality Among Female Electrical Workers in the United States', *Journal of the National Cancer Institute,* vol. 86, no. 12, pp. 921–25, 15 June 1994.

Loscher, W. and Kas, G., 'Extraordinary Behavior Disorders in Cows in Proximity to Transmission Stations', *Der Praktische Tierarz,* vol. 79, pp. 437–44, 1998.

Loscher, W. and Mevissen, M., 'Linear Relationship between Flux Density and Tumor Co-promoting Effect of Prolonged Magnetic Field Exposure in a Breast Cancer Model', *Cancer Lett.,* vol. 96, no. 2, pp. 175–80, 25 Sept. 1995.

Lotz, W.G. et al., 'Occupational Exposure of Police Officers to Microwave Radiation from Traffic Radar Devices', National Technical Information Service, Publication No. PB95-261350, June 1995.

Low Frequency Electric and Magnetic Fields: the precautionary principle for national authorities, 1996, Sweden.

Marino, A.A. et al., 'Low-level EMFs are Transduced Like Other Stimuli', *J. Neurol., Sci.* vol. 144, nos 1–2, pp. 99–106, Dec. 1996.

Matanoski, G.M. et al., 'Leukemia in Telephone Linemen', *Am. J. Epidemiol.,* vol. 137, no. 6, pp. 609–19., 15 Mar. 1993.

McIvor, M., *PACE,* vol. 21, pp. 1847–61, Oct. 1998.

McKenzie, David, presentation to the RF Spectrum Conference, Coogee, 23 Mar. 2001.

McNair, A.G.B., 1996 'Electromagnetic Energy: Summary of Community Awareness Survey' in *Inquiry into Electro-magnetic Radiation,* vol. 9, June 2000, p.1928.

Melik, G., *Magnetic Field Mitigation to Reduce VDU Interference*, ESAA, Melbourne, 1996.

Mevissen, M. et al., 'Exposure of DMBA-treated Female Rats in a 50 Hz, 50 microTesla Magnetic Field: Effects on Mammary Tumour Growth, Melatonin Levels, and T Lymphocyte Activation', *Carcinogenesis,* vol. 17, no. 5, pp. 903–10, 1996.

Michaelis, J. et al., 'Childhood Leukemia and Electromagnetic Fields: Results of a Population-based Case-control Study in Germany', *Cancer Causes Control,* vol. 8, no. 2, pp. 167–74, Mar. 1997.

Microwave News, a bi-monthly US newsletter, PO Box 1799 Grand Central Station, New York, NY 10163 USA.

Milham, S., 'Mortality in Workers Exposed to Electromagnetic Fields', *Environ. Health Perspect.*, vol. 62, pp. 297–300, Oct. 1985.

——. *Am. J. Ind. Med.*, 30, 6, pp. 702–4, 1996.

——. 'Increased Incidence of Cancer in a Cohort of Office Workers Exposed to Strong Magnetic Fields', *Am. J. Ind. Med.,* vol. 30, no. 6, pp. 702–4, Dec, 1996.

——. 'Carcinogenicity of Electromagnetic Fields', *European Journal of Oncology,* vol. 3, pp. 93–100, 1998.

Milham, S. and Ossiander, E.M., 'Historical Evidence that Residential Electrification Caused the Emergence of the Childhood Leukemia Peak', *Medical Hypotheses*, vol. 56, no. 3, pp. 290–5, 2001.

Milham, S. et al., 'Magnetic Fields From Steel-belted Radial Tires: Implications for Epidemiologic Stucies', *Bioelectromagnetics,* vol. 20, no. 7, pp. 440–5, Oct. 1999.

Miller, A.B. et al., 'Leukemia Following Occupational Exposure to 60-Hz Electric and Magnetic Fields Among Ontario Electric Utility Workers', *Am. J. Epidemiol.,* vol. 144, no. 2, pp. 150–60, 15 July 1996.

Muscat, J, 'Epidemiological Study of Cellular Telephone Use and Malignant Brain Tumors', Presented at WTR Second State of the Science Colloquium in Long Beach, California, 1999.

NHMRC, 'Interim Guidelines on Limits of Exposure to 50/60 Hz Electric and Magnetic Fields (1989)', *Radiation Health Series 30*; Published by ARL on behalf of NHMRC, 1989

——. *Procedures for Testing Microwave Leakage From Microwave Ovens*, (1985).

National Institute of Environmental Health Sciences and US Department of Energy, *Questions and Answers About EMF*, US Government Printing Office, Washington DC, 1995.

National Institute of Environmental Health Sciences, 'Questions and Answers: EMF in the workplace', www.niehs.nih.gov/emfrapid/html/Q&A-Workplace.html.

National Institute of Environmental Health Sciences and National Institute of Health, *Report on Health Effects from Exposure to Power-Line Frequency Electric and Magnetic Fields*, 1999.

National Radiation Protection Board (NRPB), *ELF Electromagnetic Fields and the Risk of Cancer*, Documents of the NRPB, vol. 12, no. 1, 2001.

National Research Council, *Possible Health Effects of Exposure to Residential Electric and Magnetic Fields,* National Academy Press, Washington, DC, 1996.

New York State Power Line Project, First recommendation of the Final Report, 1 July 1987.

Nicholas, J.S. et al., 'Magnetic Field Exposure to Airline Pilots', NIEHS, RAPID project abstract 23, *The Annual Review of Research on Biological Effects of Electric and Magnetic Fields from the Generation, Delivery and Use of Electricity*, Tucson, USA, Sep. 13–17, 1998.

Nordenson, I. et al., 'Chromosomal Aberrations in Peripheral Lymphocytes of Train Engine Drivers', BEMS Annual Meeting, Munich, 9-16 June 2000.

Örtendah and Högstedt, 'Electro-smog and Chemistry', *Heavy Metal Bulletin No. 2*, 1995.

Parliament of Australia, *Inquiry into Electromagnetic Radiation*, Report of the Senate Environment, Communications, Information Technology and the Arts References Committee, May 2001.

Peach, H.G., Bonwick, W.J. and Wyse, T., *Report of the Panel on Electromagnetic Fields and Health to the Victorian Government*, Victorian Government, Melbourne, 1992 .

Pereira, C. and Edwards, M., 'Parotid Nodular Fasciitis in a Mobile Phone User', *J. Laryngol. Otol.*, vol. 114, no. 11, pp. 886–7, 2000.

Perry, F.S. et al., 'Environmental Power-frequency Magnetic Fields and Suicide', *Health Physics,* vol. 41, no. 2, pp. 267–77, Aug. 1981.

Perry, S. et al., 'Power Frequency Magnetic Field; Depressive Illness and Myocardial Infarction', *Public Health,* vol. 103, no. 3, pp. 177–80, May 1989.

Perry, S. and Pearl, L., 'Power Frequency Magnetic Field and Illness in Multi-storey Blocks', *Public Health,* vol. 102, no. 1, pp. 11–18, 1988.

Persson, B.R.R. et al., 'Blood-brain Barrier Permeability in Rats Exposed to Electromagnetic Fields Used in Wireless Communication', *Wireless Network,* vol. 3, pp. 455–61, 1997.

Phillips, J.L. et al., 'Transferrin Binding to Two Human Colon Carcinoma Cell Lines: Characterization and Effect of 60-Hz Electromagnetic Fields', *Cancer Research,* vol. 46, no.1, pp. 239–44, Jan. 1986.

Poole, C. et al., 'Depressive Symptoms and Headaches in Relation to Proximity of Residence to an Alternating-current Transmission Line Right-of-way', *Am. J. Epidemiol.* vol. 137, no. 3, pp. 318–30, 1 Feb. 1993.

Prato, F.S. et al., 'Blood-brain Barrier Permeability in Rats is Altered by Exposure to Magnetic Fields Associated with Magnetic Resonance Imaging at 1.5 T.', *Microsc. Res. Tech.,* vol. 27, no. 6, pp. 528–34, 15 Apr. 1994.

Preston-Martin, S. et al., 'Risk Factors for Gliomas and Meningiomas in Males in Los Angeles County', *Cancer Research,* vol. 49, no. 21, pp. 6137–43, 1 Nov. 1989.

Pullman, C. and Szymanski., S., *Electromagnetic Fields (EMFs) A Training Workbook for Working People*, Labor Institute, NYC, 1995.

Quan, R. et al., 'Effects of Microwave Radiation on Anti-infective Factors in Human Milk', *Pediatrics,* vol. 89 no. 4 (pt. 1), pp. 667–9, Apr., 1992.

Redelmeier, D.A. and Tibshirani, R.J., 'Association Between Cellular-telephone Calls and Motor Vehicle Collisions', *New. Eng. J. Med.,* vol. 336, no. 7, pp. 453–8, 13 Feb. 1997.

Reichmanis, M. et al., 'Relation Between Suicide and the Electromagnetic Field of Overhead Power Lines', *Physiol. Chem. Physics,* vol. 11, no. 5, pp. 395–403, 1979.

Reiter, R.J., 'Electromagnetic Fields and Melatonin Production', *Biomed. Pharmacother.,* vol. 47, no. 10, pp. 439–44, 1993.

Reiter, R.J. et al., 'A Review of the Evidence Supporting Melatonin's Role as an Antioxidant', *J. Pineal Res.,* vol. 18, no. 1, pp. 1–11, Jan. 1995.

Reiter, R.J. and Robinson, J., *Melatonin: Your Body's Natural Wonder Drug*, Bantam, New York, 1995.

Rodvall, Y. et al., 'Occupational Exposure to Magnetic Fields and Brain Tumours in Central Sweden', *Eur. J. Epidemiol.,* vol. 14, no. 6, pp. 563–9, Sept., 1998.

Roth, J.A. et al., 'Melatonin Promotes Osteoblast Differentiation and Bone Formation', *J. Biol. Chem.,* vol. 274, no. 31, p. 22041–7, July 1999.

Rubin, P. et al., 'Disruption of the Blood-brain Barrier as the Primary Effect of CNS Irradiation', *Radiother. Oncol.,* vol. 31, no. 1 pp. 51–60, Apr. 1994.

Salford, L.G. et al., 'Permeability of the Blood-brain Barrier Induced by 915 MHz

Electromagnetic Radiation, Continuous Wave and Modulated at 8, 16, 50 and 200 Hz', *Microsc. Res. Tech.*, vol. 27, no. 6, pp. 535–42, Apr. 1994.

Salzburg Resolution on Mobile Telecommunication Base Stations, International Conference on Cell Tower Siting Linking Science & Public Health, Salzburg, 7–8 June 2000.

Sandstrom, M. et al., 'Neurophysiological Effects of Flickering Light in Patients with Perceived Electrical Hypersensitivity', *J. Occup. Environ. Med.*, vol. 39, no. 1, pp. 15–22, Jan. 1997.

Santucci, P. et al., *New Eng. J. Med.*, 339 pp. 1371–4, 5 Nov. 1998.

Sastre, A et al., 'Nocturnal Exposure to Intermittent 60 Hz Magnetic Fields Alters Human Cardiac Rhythm', *Bioelectromagnetics,* vol. 19, no. 2, pp. 98–106, 1998.

Sastre, A., 'Susceptibility of Implanted Pacemakers and Defibrillators to Interference by Power-Frequency Electric and Magnetic Fields', EPRI 1998.

Savitz, D.A. et al., 'Magnetic Field Exposure from Electric Appliances and Childhood Cancer', *Am. J. Epidemiol.,* vol. 131, no. 5, pp. 763–73, May 1990.

——. 'Magnetic Field Exposure and Cardiovascular Disease Mortality Among Electric Utility Workers', *Am. J. Epidemiol.,* vol. 149, no. 2, pp. 135–42, 15 Jan. 1999.

——. 'Case-control Study of Childhood Cancer and Exposure to 60-Hz Magnetic Fields', *Am. J. Epidemiol,.* vol. 128, no. 1, pp. 21–38, July 1988.

——. 'Electrical Occupations and Neurodegenerative Disease: analysis of US Mortality Data', *Arch. Environ. Health,* vol. 53, no. 1, pp. 71–4, Jan.-Feb. 1998.

Schirmacher, A. et al., 'Electromagnetic Fields (1.8 GHz) Increase the Permeability to Sucrose of the Blood-brain Barrier in vitro', *Bioelectromagnetics,* vol. 21, no. 5, pp. 338–45, July 2000.

Schüz, J. et al., 'Residential Magnetic Fields as a Risk Factor for Childhood Acute Leukaemia: Results from a German Population-based Case-control Study', *Int. J. of Cancer*, vol. 91, no. 5, pp. 728–35, Mar. 2001.

Sobel, E. et al., 'Occupations with Exposure to Electromagnetic Fields: A Possible Risk Factor for Alzheimer's Disease', *Am. J. Epidemiol.* vol. 142, no. 5, pp. 515–24, 1 Sep. 1995.

——.'Elevated Risk of Alzheimer's Disease Among Workers with Likely

Electromagnetic Field Exposure', *Neurology,* vol. 47, no. 6, pp. 1477–81, Dec. 1996.

Speers, M.A. et al., 'Occupational Exposures and Brain Cancer Mortality: a Preliminary Study of East Texas Residents', *Am. J. Ind. Med.,* vol. 13, no. 6, pp. 629–38, 1988.

Spitz, M.R. and Johnson, C.C., 'Neuroblastoma and Paternal Occupation. A Case-control Analysis', *Am. J. Epidemiol.,* vol. 121, no. 6, pp. 924–9, June 1985.

Sridhar, S. and McIvor, M. *New Eng. J. Med.*, 339 pp. 1394–5 letter, 5 Nov. 1998.

Stenberg, B., 'The Sick Building Syndrome (SBS) in Office Workers. A Case-referent Study of Personal, Psychosocial and Building-related Risk Indicators', *Int. J. Epidemiol.*, vol. 23, no. 6, pp. 1190–7, Dec. 1994.

Stevens, R.G., Wilson, B.W. and Anderson, L.E. (eds.), *The Melatonin Hypothesis: Breast Cancer and Use of Electric Power*, Battelle, 1997.

Swedish Trade Union confederation LOA, 'Cancer and Magnetic Fields at the Workplace.' Stockholm: LOA, 1993. From *Electrosensitivity News*, vol. 3, no. 3 p .8.

Swedish Union of Clerical and Technical Employees in Industry (SIF),
– *Hypersensitive in IT Environments*, Stockholm, 1996.
– *Noll Risk,* Stockholm, 1999.

Swiss Re, *Electrosmog—a phantom risk*, 1996.

Theriault, G. et al., 'Cancer Risks Associated with Occupational Exposure to Magnetic Fields Among Electric Utility Workers in Ontario and Quebec, Canada, and France: 1970-89', *Am. J. Epidemiol.,* vol. 139, no. 6, pp. 550–72, 15 Mar.1994.

Thomas, J.R. et al., 'Low-intensity Magnetic Fields Alter Operant Behavior in Rats', *Bioelectromagnetics,* vol. 7, no. 4, pp. 349–57, 1986.

Tomenius, L., '50-Hz Electromagnetic Environment and the Incidence of Childhood Tumors in Stockholm County', *Bioelectromagnetics,* vol. 7, no. 2, pp. 191–207, 1986.

Tornqvist, S., 'Paternal Work in the Power Industry: Effects on Children at Delivery', *J. Occup. Environ. Med.,* vol. 40, no. 2, pp. 111–7, Feb. 1998.

Trimmel, M. and Schweiger, E., 'Effects of an ELF (50Hz, 1 mT) Electromagnetic Field (EMF) on Concentration in Visual Attention, Perception and Memory Including Effects of EMF Sensitivity', *Toxicol Lett.*, 96–7; pp. 377–82, Aug. 1998.

Tynes, T. et al., 'Incidence of Breast Cancer in Norwegian Female Radio and

Telegraph Operators', *Cancer Causes Control* vol. 7, no. 2, pp. 197–204, Mar. 1996.

University of Melbourne Statistical Consulting Centre, *Epidemiological Studies of Cancer and Powerline Frequency Electromagnetic Fields, A Meta-analysis*, Report No. 242, pp. 103–4, Dec. 1990.

US Department of Health and Human Services, *NIOSH Publications on Video Display Terminals*, 3rd Edition, Sept. 1999, p. 55.

US EPA, *An Evaluation of the Potential Carcinogenicity of Electromagnetic Fields*, US Government Printing Office, Washington, 1990.

US EPA, *Electric Magnetic Fields in Your Environment*, US Government Printing Office, Washington, 1992.

Vedholm, K. and Hamnerius, Y.K., 'Personal Exposure from Low Frequency Electromagnetic Fields in Automobiles', 2nd World Congress for Electricity and Magnetism in Biology and Medicine, Abstract F-9, Bologna, Italy, 1997.

Vignati, M. and Giuliani, L., 'Radiofrequency Exposure Near High-voltage Lines', *Environmental Health Perspectives*, 105, Supplement 6: Radiation and Human Health pp. 1569–73, Dec. 1997.

Villeneuve, P.J. et al., L., 'Non-Hodgkin's Lymphoma Among Electric Utility Workers in Ontario: the Evaluation of Alternate Indices of Exposure to 60 Hz Electric and Magnetic Fields', *Occup. Environ. Med.*, vol. 57, no. 4, pp. 249–57, Apr. 2000.

Watanabe, F., *Journal of Agricultural and Food Chemistry,* vol. 46, pp. 206–210, 1998.

Watt, D.G., 'Mobile Telephones and Lesions of the Mouth', *Br. Dent. J.*, vol. 189, no. 5, pp. 237, 2000.

Wertheimer, N. and Leeper, E., 'Electrical Wiring Configurations and Childhood Cancer', *Am. J. Epidemiol.*, vol. 109, no. 3, pp. 273–84, Mar. 1979.

Wertheimer, N. et al., 'Childhood Cancer In Relation to Indicators of Magnetic Fields From Ground Current Sources', *Bioelectromagnetics*, vol. 16, no. 2, pp. 86–96, 1995.

Whittemore, T.R. et al., 'Characterization of Magnetic Fields in Electric Vehicles and Comparison to Those in Conventional Vehicles', NIEHS, RAPID project, abstract 21, The Annual Review of Research on Biological Effects of Electric and Magnetic Fields from the Generation, Delivery and Use of Electricity, Tucson, USA, 13–17 Sep. 1998.

World Health Organisation, 'Non-Thermal Effects of RF Electromagnetic Fields',

Proceedings of International Seminar on Biological Effects of Non-Thermal Pulsed and Amplitude Modulated RF Electromagnetic Fields and Related Health Risks, Munich, 20–21 Nov. 1996.

——. *Electromagnetic Fields and Public Health Cautionary Policies*, Mar. 2000.

Youbicier-Simo, B.J. et al, *Progress in Radiation Protection*, vol. 1, pp. 218–23, 1999.

Zaffanella, L., US Dept. of Energy's 'EMF Engineering Review Symposium', Charleston, 29 Apr. 1998.

——. 'General Public Exposure. Results of the 1,000 People Survey', RAPID project abstract 25, The Annual Review of Research on Biological Effects of Electric and Magnetic Fields from the Generation, Delivery and Use of Electricity, Tucson, USA, 13–17 Sep. 1998.

Zaret, M. et al., 'Cataract After Exposure to Non-ionizing Radiant Energy', *British Journal Ophthalmol.*, vol. 60, no. 9, pp. 632–7, Sep. 1976.

Zipes, D., *Circulation*, vol. 100, pp. 387–92, 27 July 1999.

Recommended reading

Becker, R.O. and Seldon, G., *The Body Electric*, William Morrow, New York, 1985.

Becker, R.O. and Seldon, G., *Cross Currents*, Bloomsbury, London, 1991.

Brodeur, P., *Currents of Death,* Simon and Schuster, New York, 1989.

Cellular Phone Taskforce, *No Place to Hide,* newsletter, PO Box 100404, Brooklyn, New York, NY 11210.

Cherry, N., *Criticism of the Health Assessment in the ICNIRP Guidelines for Radiofrequency and Microwave Radiation (100kHz—300 GHz)*, Lincoln University, New Zealand, 22 Oct. 1999.

Dalton, L., *Radiation Exposures*, Scribe, Victoria, 1991.

EMRAA, *EMRAA News,* a quarterly newsletter on electromagnetic radiation, PO Box 589, Sutherland 1499, Australia.

Grant, L., *The Electrical Sensitivity Handbook—How Electromagnetic Fields (EMFs) Are Making People Sick*, Weldon Publishing, Arizona,1995.

Grant, L, *Workstation Radiation*, Weldon Publishing, Arizona, 1992.

Lai, H., 'Neurological Effects of Radiofrequency Electromagnetic Radiation Relating to Wireless Communication Technology', paper presented at IBC-UK Conference, *Mobile Phones—Is there a Health Risk,* Brussels, Belgium, 16–17 Sep. 1997.

Lai, H., 'Neurological Effects of Radiofrequency Electromagnetic Radiation, paper presented to the Workshop on Possible Biological and Health Effects of RF Electromagnetic Fields', Vienna, Austria, 25–28 Oct. 1998.

Microwave News, a bi-monthly US newsletter, PO Box 1799 Grand Central Station, New York.

Pullman, C. and Szymanski, S., *Electromagnetic Fields (EMFs) A Training Workbook for Working People*, Labor Institute, New York, 1995.

Swedish Union of Clerical and Technical Employees in Industry (SIF), *Hypersensitive in IT Environments*, Stockholm, 1996.

Index